T0092951

THE LAST OF ITS KIND

THE LAST OF
ITS KIND

THE SEARCH FOR THE GREAT AUK
AND THE DISCOVERY OF
EXTINCTION

GÍSLI PÁLSSON

PRINCETON UNIVERSITY PRESS

PRINCETON & OXFORD

Published by Princeton University Press
41 William Street, Princeton, New Jersey 08540
99 Banbury Road, Oxford OX2 6JX

press.princeton.edu

All Rights Reserved

ISBN 978-0-691-23098-6
ISBN (e-book) 978-0-691-23099-3

British Library Cataloging-in-Publication Data is available

Editorial: Alison Kalett and Hallie Schaeffer
Production Editorial: Jill Harris
Text Design: Karl Spurzem
Jacket Design: Karl Spurzem
Production: Danielle Amatucci
Publicity: Matthew Taylor and Kate Farquhar-Thomson
Copyeditor: Jennifer Harris

Jacket image: Iconographia Zoologica, *Chenalopex impennis*, between 1700 and 1880, Special Collections of the University of Amsterdam

This book has been composed in Arno pro with Trade Gothic Next LT Std

Printed on acid-free paper. ∞

Printed in the United States of America

10 9 8 7 6 5 4 3 2 1

For Guðný

The [Great Auk] may . . . be described, for it is little known to People, and Nature has deprived it of that which it has bestowed on other Birds, that is to say Wings; yet it flies with its Wings in the Sea as swiftly as flying Birds in the Air.

—Guðni Sigurðsson (1714–80), 1770, *Gare-Fowl Books*, 1858

We should regard every living being . . . as a going on in the world. . . . Thus whether we are speaking of human or other animals, they are at any moment what they have become, and what they have become depends on whom they are with.

—Tim Ingold (b. 1948), *Imagining for Real*, 2022

I call this era of man-made mass death a time of "double death." . . . It is not that the whole biosphere is overtaken with death, but rather that death continues to pile up; renewal and resilience cannot keep the pace.

—Deborah Bird Rose (1946–2018), *Shimmer*, 2022

CONTENTS

FIGURE 0.1. Ole Worm's great auk, Copenhagen.
(Wikimedia Commons.)

I wonder, can such a process possibly be catalogued or catego-
rized, given the speed of change and the complexities involved—
and what would be the point?

As I walked to the museum in Copenhagen, I observed gulls
flying overhead, their raucous cries drowning out the noise of
the city—as if they were asking me to take their greetings to the

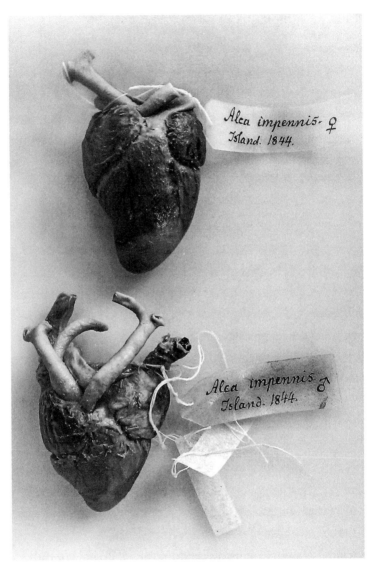

FIGURE 0.2. The last great auk hearts. (Courtesy of the Natural History Museum, Copenhagen.)

taxidermied birds inside. Another day in downtown Copenhagen, I encountered a crowd of young people who were loudly marching, brandishing placards with the slogans of a new international movement: Extinction Rebellion. Most noticeable were the placards that read: *There is no Planet B!*

For decades, the great auk had held the place of honor in what was then Copenhagen's Zoological Museum, ornithologist and former collections manager Jan Bolding Kristensen told me (see plate 1). In his office was a life-size great auk made of plastic, wearing sunglasses and a necktie, as if to indicate that it was part of the museum staff. In the 1980s, Kristensen made a visit to Iceland, sailing past Eldey, where the birds whose viscera are preserved in the museum's great glass jars had once laid and incubated their beautiful eggs.

Kristensen showed me the museum's old logo: it depicted a great auk. Sadly, the logo was replaced following the reorganization of the natural history collections in Copenhagen. The great auk was no longer regarded as being representative of the several museums and their diverse fields. Before, when the zoological museum celebrated its foundation day, the public had been invited to come and see the logo's inspiration for themselves. Hundreds of people would walk around the taxidermied great auk, that fearless-looking bird with the great beak; they would think about it and put questions to the staff. Birds that no longer exist had, and still have, a special attraction.[3] They have much to teach us.

The *Gare-Fowl Books*

I never saw a great auk when growing up in Iceland, where they had once been quite common. What first brought my serious attention to the fate of the species and its continued relevance

FIGURE 0.3. Logo of the old Zoological Museum, Copenhagen. (Image: Jan Bolding Kristensen.)

was Elizabeth Kolbert's 2014 book, *The Sixth Extinction*, which has a chapter on the "original penguin." I was also struck by a 2018 article by Petra Tjitske Kalshoven, "Piecing Together the Extinct Great Auk," which cited in some detail a group of intriguing manuscripts—known as the *Gare-Fowl Books*—describing an

expedition to Iceland by the nineteenth-century British naturalists John Wolley and Alfred Newton, "gare-fowl" being an old
name for the great auk. Kolbert mentioned these manuscripts,
I realized. So did Errol Fuller and Jeremy Gaskell in their books
on the great auk from 1999 and 2000. Last, a 2013 article by
historian of science Henry M. Cowles, "A Victorian Extinction,"
drew my attention to the Victorian concerns with evolution and
animal protection that informed Alfred Newton's groundbreaking work on the concept of extinction.

Like their contemporaries, Wolley and Newton busily collected birds' eggs and specimens, classifying and recording
them in the fashion of the Victorian age. When they set off for
Iceland in 1858, they hoped to visit Eldey and study the rare
great auk. They hoped to observe the bird's behavior and habits
and, perhaps, bring home an egg or a skin or a stuffed bird or two
for their own cabinets of curiosities—unaware of the fact that
the species had already been hunted to extinction. When they
left Victorian England for Iceland, they teased that this was a
"genuinely awkward expedition." And so it proved to be, in
many ways. They never made it to Eldey. Like me, they never
saw a great auk on Iceland, not even stuffed.

Prior to the killing of the last great auks, extinction was either
seen as an impossibility or trivialized as a "natural" thing. The
great taxonomist Carl von Linné, or Linnaeus (1707–78),
imagined that a living species could never disappear; for evolutionary theorist Charles Darwin (1809–82), species would
naturally come and go in the long history of life. The great auk
brought home the fact that a species could perish quite quickly
and, moreover, not naturally, but primarily as a result of human
activities. No other extinction had been documented as carefully.

During their historic expedition to Iceland in 1858, Wolley
and Newton collected impressions of great auk hunting,

through substantial interviews with the men who took part in the latest hunts and the women who skinned and mounted the birds, along with their prices and sales on foreign markets to collectors of "curiosities." These impressions were preserved in the set of five handwritten notebooks Wolley titled the *Gare-Fowl Books* (see plate 2).[4] Now archived in Cambridge University Library in England, their hundreds of pages are written in several languages (English, Icelandic, Danish, and German). As an anthropologist and an Icelander, once I had seen the *Gare-Fowl Books*, there was no return: I had to dive into the text and visit zoological museums and archives—and to write this book. For me, the great auk opened wide an intellectual window into ideas of extinction and their relevance to the current mass disappearance of species.

Now that I know the great auk's long history, I feel as if the stuffed birds in the Copenhagen museum were once my neighbors or acquaintances. As a scientist, I know that their viscera are stored in alcohol to preserve them and to enable people to study them. But I wonder if the organs are in a constant state of inebriation from the alcohol, existing beyond the bounds of real time, in a sort of euphoric oblivion? Generations of visitors, of all ages and many nationalities, have passed by these jars of preserved bird parts over the past century and a half. What observations did they take home? The hearts stored in one jar are no longer beating, but no doubt many visitors on my side of the glass have wondered, as I do, how they would have pulsed when the bird's blood was still flowing—and whether they could be resuscitated, by electric shock or genetic reconstruction. The eyes of the last male great auk are kept in another jar. I see them staring, gazing both into the past and into my own eyes.

In the Field

Bird-watching was part of my childhood in Iceland. Is that why
the fate of the great auk attracts me so? As a youngster, from
the age of ten, I collected birds' eggs. Attending the cows in the
meadows and passing through nearby wetlands gave me an opportunity to explore the rich and noisy, sometimes deafening,
birdlife during the peak of the nesting season. I found the lively
birds a striking contrast to the lazy and stubborn cows. I blew
out the contents of the eggs I found, and kept the unbroken
ones cushioned in hay in a cardboard box. My largest was from
a swan (*Cygnus cygnus*)—nearly as big as the egg of the great
auk. It's remarkable, I think, that the same term is applied to the
music of the swans as to the final performance of a living being:
a "swansong," the metaphor based on the ancient folk belief that
a swan's final song is its most beautiful.

Perhaps my interest in the great auk began, instead, when I
did anthropological fieldwork in the Icelandic town of Sandgerði (from 1979 to 1981), about ten miles northwest of Keflavík
International Airport; Wolley and Newton had worked in Sandgerði over a hundred and twenty years earlier, interviewing
some of the men and women who hunted, killed, and processed
great auks. I was conducting a similar study, seeking information on skippers reputed to be "good fishers." I remember walking from one farmhouse to another, and having long talks with
men who knew the fishing grounds like the backs of their
hands—and a younger generation of skippers who fished the
same grounds under entirely different conditions, using sophis-
ticated electronic equipment. I drank a lot of coffee, and there
was some intensive smoking. Some days, I would drive out of
town past Keflavík International Airport and all the way to the
southern end of the peninsula, the Reykjanes "toe," looking at

old farmsteads and fishing stations and striving to grasp what life had been like in the days of the rowing-boat fishery. I had not known then about the two British scientists who rode this way in 1858 to learn about the great auk hunts.

When it dawns upon me that it is Sandgerði that directly connects me to the great auk expedition of 1858, I dig out the tape recordings I made during that field trip in 1979—a total of ten tapes. I ask around for a machine that will play them (Sony UX-S90, now a rarity), and sit down to listen. The sound quality is good, and my interview subjects express themselves clearly, but their voices seem to come from the distant past, speaking of a long-vanished world. The boat foremen explain what their role was: the importance of sitting in the proper place onboard, how the boat was launched from the beach, how the fishing trips were carried out, what it meant to be a "good fisher," and how shares of the value of the catch, or the catch itself, were divided among the crew. My informants sound like they were living in the Middle Ages.

During my first stay in Sandgerði in 1979, I was offered the opportunity to go out on a fishing boat, laying nets in the Reykjanes fishing grounds near Eldey—in the dangerous straits Wolley and Newton never ventured into, notorious for crosscurrents and rough seas. One could not say no to such a challenge. The forecast was fine, but soon after we departed, the weather turned. When I came ashore more than twenty-four hours later, head spinning and still wobbly on my legs, with a taste of vomit in my mouth, the humorists in the crew joked that I had given the birds plenty to feed on. As an anthropologist, I find it interesting now that the fishers used the recovery from seasickness (*sjóveiki*) metaphorically for the process of learning. I remember the pleasant feeling near the end of the fishing trip, as I regained my balance: I began enjoying the boat's movements and could attune to the tasks at hand. This is referred to as *að sjóast* in Icelandic, "to become the sea"—similar to, but not quite the same as, the English phrase "finding one's sea legs." The Icelandic metaphor makes good sense to me. As we become skillful at our tasks, we are not simply internalizing mental scripts, what one needs to "know," in the conventional sense. More importantly, the bodily sensation of nausea gives way to skillful grounding—it's literally a "gut reaction." As we immerse ourselves in our activities, engaging with the immediate context of learning, we begin to feel at home in both our bodies and in the world.[5]

In the evenings, during my field seasons in Sandgerði, I would make notes of what I had learned about the battle of these people who go out and catch fish. My notebook proved useful later, although it is a small one. I am filled with admiration for Wolley and Newton, who had nothing to rely on but pencil, pen, and paper. Unlike them, I had a camera and a tape recorder—and the benefit of being a fisherman's son and a native speaker of Icelandic. I benefited, too, from a brand-new electronic gadget fishermen would call "the spy." It automatically searched for dialogues at sea on the inter-boat radio, allowing the fieldworker (and skippers) to pick up endless gossip about fishing places, gear, weather, and personalities. Sometimes complicated codes and euphemisms were used, given the competitive nature of fishing; these I needed to translate and interpret. I recorded hours of such dialogues, and they became an extremely valuable source of anthropological insights, somewhat like the *Gare-Fowl Books* for Wolley and Newton.

My father, who was born in East Iceland, became one of my informants. I was surprised to learn that he had begun his fishing career on a small motorboat at Sandgerði in 1939, at the age of seventeen. My mother was relieved when he quit fishing. He decided to take a job in a fish plant when I was born, ten years later; I was their second child in just one year. I can understand her apprehensions. My maternal grandfather was the foreman of an open rowing boat in the early twentieth century. For as long as I can remember, a large photograph hung on the living-room wall at home, showing his boat and crew on their way out to fish. The men were posed between the thwarts in their oil-skins. They look serious. Their safety at sea could never be taken for granted; the boat had no motor, only oars, and the nearest landing place on the mainland was far away. The south shore of Iceland is an endless expanse of sands with hardly any harbors or safe havens, often whipped by crashing surf.

In my father's time, the end of the fishing season was May 11, when accounts were settled and fishers moved on to other tasks; the choice of date adapted to fish migrations, ecological cycles, and social life. Fishing stocks were seen then as an infinite resource and were open to all (with minor limitations regarding breeding grounds), as long as one had a boat and a crew. Before long, though, the fish populations began to decline. Cod, the main species fished around Iceland, might not have become extinct, but catches were unreliable. A couple of years after my fieldwork, in 1983, the Icelandic Parliament decided to launch a quota system, radically redefining fishing: it was now the responsibility of humans to prevent collapses of fish populations. Soon, the ocean would come to be seen as the ultimate aquarium, inviting fundamental questions of inside versus outside, the observer and the observed—much the same questions as were raised in the nineteenth century by the extinction of the great auk.

Historic Sites

At some points, while writing this book, I have felt the need to mentally relive Wolley and Newton's journey through visiting the key Icelandic sites. One autumn day in 2018, for example, I

go back to Sandgerði—the place that, above all, links me to
their quest for the great auk. Having located the remains of the
old cottage from which the town receives its name, I drive ten
miles south past the airport to Kirkjuvogur, to see the manor
house where the two British naturalists lived. The cluster of
buildings that stood there in 1858 are gone. I knock at the door
of a modern house that stands about where I think the old farmstead was located and am told with a smile (perhaps they are
used to the question?) that the present residents know nothing
of the great auk hunters. The buildings on the farmstead of Kotvogur, closer to the sea, are half-collapsed. The ruins of the turf
structures that Wolley and Newton visited are scattered around,
and the impressive local cemetery recalls that this was an
important and busy place in the nineteenth century.

I try to identify the landing place where the foreman recited
the fisher's prayer before launching the boat in search of great
auks, and where the British visitors clung to the hope of getting
out to sea themselves and reaching the island of Eldey—
although the weather never lifted, and they never landed on the
island to learn for themselves that the gare-fowl was gone. A
little farther south, I decide to look for the cairn that Wolley had
built, a drystone cone of rocks, probably to serve as a
landmark—and a monument to their expedition. I follow the
trail that he and Newton must have taken along the cliffs; it
takes me considerable driving and walking to reach their observation point. Over the centuries, this coastline has seen serious
erosion due to high seas. One of the worst floods recorded occurred in the middle of a January night in 1799, ruining an entire
hamlet; one woman drowned and many boats were shattered.
A nearby church was blown away. Equipped with Wolley's
drawing, depicting the cairn with Eldey in the distance, I finally
find the spot, about five miles south of Kirkjuvogur, where the

FIGURE 0.4. Todd McGrain's great auk on Reykjanes, looking toward Eldey. (Photo: Gísli Pálsson.)

grassland meets the ocean. Wolley's drawing is clearly accurate, giving a good idea of the contours of the landscape. I can sense his excitement as he built his cairn, Eldey over his shoulder as he worked. But the cairn itself—Wolley's monument—is gone.

I continue south along the coast to the westernmost point of the peninsula. Here stands a tall statue of a great auk by US artist Todd McGrain, part of his Lost Bird Project, memorials of extinctions at historical sites. It was installed by permission of the land-owners, descendants of a man who had rowed out to Eldey on one of the last hunting trips. The sculpture is a pleasant surprise, a proud bird, about my height, gazing mournfully out to sea. Putting my arm around the bird, I look out in the same direction, and catch a glimpse of Eldey through the drizzle. The waves crash on the cliffs like a reminder of those old expeditions to the island that had such a fateful impact on the lives of both birds and humans—and on the modern framing of extinction.

THE LAST OF ITS KIND

CHAPTER 1

THE ROAD TO EXTINCTION

In 1858, when John Wolley and Alfred Newton set off from Britain on their expedition to Iceland, the great auk (*Pinguinus impennis*) was reported to be in serious decline. Wolley's friend William Proctor, keeper of the bird collection at Durham University, had traveled to Iceland in 1833 and 1837, partly in order to seek out great auks. Proctor told Wolley that sightings were now rare in Iceland and that he had not seen any of the birds.[1] Wolley's fellow-student William Milner, who went to Iceland some years later, inquired about great auks on his travels and was informed that none had been seen recently, though two had been caught two years earlier, in 1844. Milner's account of his visit gave rise to a strong suspicion that the species was not only rare but vanishing.

Wolley took a keen interest in discussions of rare birds, and he resolved to go to Iceland with the same intention as his friends. He invited Alfred Newton, then making a name for himself as a zoologist at Cambridge University, to join him. Wolley and Newton met for the first time in Cambridge one October day in 1851, although they had corresponded for several years. Wolley had recently passed his final examinations in medicine at the University of Edinburgh, with excellent results, but he had decided not to pursue a career as a doctor; instead, he would follow his vocation to create a great and systematic egg collection. It would eventually become one of the largest collections ever known, numbering at least ten thousand eggs. Perhaps in Iceland Wolley could acquire an egg of the rare great auk?

Newton, who was six years Wolley's junior, had also been collecting eggs since boyhood and had kept meticulous records of the comings and goings of migratory birds. The animated letters that Wolley wrote to him during a collecting expedition in northern Scandinavia captured the younger man's imagination and stimulated his interest in nature. To Newton, the north was an "ornithological paradise," where rare bird species were nevertheless still to be found and the diversity of species was immense.[2] He and Wolley agreed that they must go to Iceland as soon as possible and seek out the great auks—although for the next seven years nothing came of their plans. They were, frankly, obsessed with finding the bird they knew by the name of "gare-fowl." When they finally set off, in 1858, their ambition was to learn as much as possible about the species during a two-month stay, during which they would visit the great auk breeding grounds on Eldey, a small island off Iceland's southwest coast.

Bad weather prevented them from even attempting to row out to Eldey. Stuck ashore, with no gare-fowl to observe, they occupied themselves with identifying the crew of the latest successful great auk hunting expeditions, interviewing as many people as possible who had seen the birds. Wolley carefully preserved their accounts—along with much other information about the great auk—in the set of notebooks now known collectively as the *Gare-Fowl Books*. On the basis of what the two British naturalists learned in Iceland in 1858, Newton, who outlived his friend and preserved his legacy, would become a leading figure in discussions of a new and politically volatile scientific concept: extinction.

Was it conceivable, Wolley and Newton wondered on their return to Britain, that this sizable bird, known to collectors around the world, was in critical decline as a result of human activities? Could it be erased from the book of life altogether? Was such a thing—unnatural extinction—possible?

In the early nineteenth century, most people, both lay and learned, believed that all the species of the living world had been created once and for all, that existing organisms could not vanish, and that new species could not appear. The Creation was seen as perfect; the principal role of the natural scientist was to document, describe, and classify the species created.

Today, the concept of species is essential to our understanding of extinction, but nobody, not even scholars, talked about extinction in those terms in the early nineteenth century. Species did not disappear. There was no name for the loss of a species—particularly not for a loss that might be detected and studied in the here and now. The English noun *extinction* (from Latin *extinctionem*) had, of course, been in use since at least the fifteenth century; it meant "annihilation." The related verb *to extinguish* meant (and still means) to quench, in the context of fires, or, figuratively, to wipe out a material thing, such as a debt. Yet in the early nineteenth century, as Cambridge scholar Gillian Beer points out, the word *extinction* was primarily "linked to the history of landed families: a line becomes extinct and with it the family name and the succession of property and practices."[3]

Not until the late 1880s were *extinction* and *species* paired, and extinction became a matter of biology and governance.[4] The species that instigated this pairing was the great auk, and it was Wolley and Newton's 1858 expedition to Iceland that sparked this important conceptual development, adding the concept of unnatural extinction to modern language and thought.

The Fossil Hunters

Before unnatural extinction—the loss of a species as a result of human activities—could be understood, the idea that creatures could become extinct by any means at all needed to be accepted. Taxonomer Carl von Linné was among those to protest that such a thing was flatly impossible. "We will never believe that a species could totally vanish from the earth," he said, and his was the prevailing viewpoint.[5]

One of the key quotes from Linnaeus's work is from his *Genera plantarum* (§ 5):

> There are as many species as there were different forms produced by the Infinite Being in the beginning. Which forms afterwards produce more, but always similar forms according to inherent laws of generation; so that there are no more species now than came into being in the beginning. Hence there are as many species as there are different forms or structures of plants occurring today, setting aside those which place or accident exhibit to be a little different (varieties).[6]

It's a heavy paragraph, and Linnaeus doesn't quite say that species do not disappear, only that there are "no more species now" than originally. He allows for the role of "place or accident" in exhibiting "varieties" (an interesting nod to evolutionary theory), but the possibility of progressively fewer species—of what we now call "extinction"—was unthinkable at the time (in 1737); life-forms, it was implied, somehow remained intact since the theological big bang.

The concept of species itself had been first developed around 1680 by British naturalist John Ray (1627–1705); about half a century later, Linnaeus proposed a taxonomic system of species in his treatise *Systema Naturæ*.[7] But Linnaeus was interested solely in species that existed in his own time; the prior history of the planet was of no importance to him and had no place in his taxonomy.

Well after Linnaeus's death, German anatomist Johann Christian Rosenmüller (1771–1820) examined a set of mysterious bones that had been discovered in a cave in southern Germany in 1748. Rosenmüller concluded that the bones represented a bear unknown to science; it had once lived, he said, but then totally disappeared.[8] This was a stunning inference at a time when the idea of what now counts as extinction hardly existed.

It was, however, George Cuvier (1769–1832) at Paris's Museum of Natural History who established extinction as a historical fact through his broader study of fossilized bones. During the mayhem of the French Revolution, which questioned just about everything, he pointed out that historical animal remains discovered by geologists and collectors in rock strata at several places in the world exposed species that had disappeared. At first, his opinions were disputed. The evidence was scant, and Cuvier's conclusions irritated people who believed that the earth had a shallow history; according to the Bible, it was only about 6,000 years old.[9]

In 1812, Cuvier published a four-volume compendium in which he presented evidence of forty-nine extinct vertebrates that he had accumulated. Cuvier—a flamboyant character with a passion for history, honored by Napoleon and invited for talks in Britain—wasn't sure how to make sense of his fossils and the vanished creatures they represented; he had no faith in what would soon come to be called "evolution." But Cuvier assumed that there was a lot to be learned from fossils. Like Alfred Newton, he was more interested in the disappearance of species than in their origins. Cuvier carefully studied the anatomy of the monstrous species accessible to him, and was eager to collect more samples for his expanding museum.

For this, Cuvier, like his male peers, was dependent on fossil hunters, who did most of the dirty work of finding and digging up fossilized bones. Also, he and his colleagues often drew upon skillful women within their domestic contexts; Sophie Cuvier sketched birds for her father.[10] Arguably, the fossil hunters and the artists deserve no less credit than the collectors and museum lecturers for launching the discovery of the origins and disappearance of species. Without their skills and painstaking work, based on their practical touch and deep understanding of context and terrain, we would have no concepts of speciation: no extinction, no evolution.

One of the key fossil hunters at the time of Cuvier was Mary Anning (1799–1847). Anning lived in Lyme, near what is now known as the Jurassic Coast in Dorset in the west of England. She has been dubbed "the greatest fossilist the world ever knew."[11] Anning came from a poor family; her father was a carpenter, and the Annings lived largely by selling fossils to scientists from museums and to keepers of private cabinets of curiosities. Mary Anning went out for walks, armed with rock hammers and carrying a sack on her back, most days. Although she did not know the mechanism, ancient sedimentary rock from the seabed, with layers containing fossils, had been lifted to the surface as the tectonic plates moved, and the fossils were brought to light, eroding out of the cliff faces of the coast of Lyme Regis.

At the early age of ten or twelve (accounts vary), Anning found the fossilized remains of a marine creature that was later named *Ichthyosaur* ("Fish Lizard"). Her discovery, a collaboration with her brother Joseph, was widely discussed: *Ichthyosaur* raised new questions about the story and fate of animals. In 1823, Anning discovered a nearly intact plesiosaur (*Plesiosaurus dolichodeirus*). Cuvier famously invalidated this discovery at first, claiming the animal to be an impossibility, but when a well-known British geologist presented the evidence and defended Anning's work, Cuvier was forced to pronounce Anning's fossil a major discovery.

It later transpired that the geological relics that Anning retrieved, reconstructed, and sold to collectors were 200 million years old. Their existence confirmed that Cuvier was right.[12] Now he was able to establish that species had truly disappeared off the face of the earth, not simply moved or gone into hiding. Some creatures, it appeared, had originated in the sea and later emerged on dry land, or even took flight. This notion threw taxonomists into confusion. What did it mean to be a bird? Feathers were not exclusively a property of "birds" or an indication of "birdness": palaeontology would reveal that some dinosaur species had been feathered.[13] And some birds, including the great auk, couldn't fly. The science of life needed to be rethought.

Anning never published a scientific paper; all that remains of her writing is a number of letters, including her correspondence with scientists. She was deeply religious and struggled to reconcile her groundbreaking discoveries with her Christian faith, resisting ideas of birth and extinction. It seems, however, that in the end she accepted the idea of change. In 1833, when visited by tourists, she is reported to have said that because her fossils were found at different levels in the cliffs, the animals in question had been created and lived at different times. Later, she remarked that, judging from her observations of fossils, there

was a "connection of analogy between the Creatures of the former and present World."[14]

While Anning became a public figure and the tourist attraction of Lyme, placing her community on the radar, few scholars
made any reference to her in their writings; until recently she
received little credit for her contributions to science. Stratified
by gender and social class, science had no space for her name.
"The world has used me so unkindly," she said. "These men of
learning have sucked my brains." As American palaeontologist
Stephen Jay Gould has remarked, Anning was "probably the
most important unsung (or inadequately sung) collecting force
in the history of paleontology."[15] She paved the way for Charles
Darwin, Alfred Wallace, Alfred Newton, and many others.

It was due to the work of Anning that naturalists in the mid-
nineteenth century became accustomed to the idea that prehistoric "species" could be wiped off the face of the earth; that they
could become "extinct"—at least, in the context of fossil discoveries. Fossils were understood to represent animals that no longer existed. Yet while the people of the time debated the reasons
for the disappearance of these strange prehistoric creatures,
most assumed that the causes were "natural," in the common
sense. People referred to myths of floods and chaos, or to the
collaborative efforts of gods and humans.

Biology Is Invented

One of the successes of the nineteenth century was the discovery and calibration of deep time in 1859—only a year after Wolley
and Newton's Iceland expedition.[16] This was the "time revolution," based on the discovery by two British amateurs of hand-
held stone tools among the bones of ancient animals at Somme
in France. Humans, it was now established, had lived in deep

geologic time, much longer than the 6,000 years postulated in biblical accounts. History had begun. Charles Darwin (1809–82) and Alfred Russel Wallace (1823–1913) went on to demonstrate that the story of life itself on earth was far longer than had been believed. (A solid estimate of the age of the earth, however, was achieved only at the very end of the nineteenth century, with the discovery of radioactivity.)

Darwin and Wallace reasoned that life was a constant process of change, in which natural selection played a crucial role. Only the variants of a certain species that were "fittest" for the prevailing conditions would survive to give rise to new species, while others would decline or disappear. Darwin's first manuscripts about his theory of natural selection date from 1842 and 1844—years during which the great auk was pushed to extinction. When a species became uncommon or rare, Darwin argued, it was an indication of impending extinction. The fossils spoke for themselves. Darwin, however, decided not to publish in the 1840s, fearing that his theory would spark controversy.

Wallace developed his theory of evolution in Borneo in 1855, as he waited for the end of the monsoon season. He had observed a simple rule: each species "had for immediate antetype a closely allied species existing at time of its origin."[17] One species must follow on from another.

When Darwin received a letter from Wallace, indicating that he had reached theoretical conclusions like his own, he knew he could no longer delay the announcement of the theory of natural selection. Remarkably, he offered Wallace an opportunity to share with him the honor. The announcement of their theory was a historic moment—although few of those present at the meeting of the Linnean Society of London on 1 July 1858 realized its significance. Newton, who was then in Iceland along with his friend Wolley, frustrated by their inability to find a great auk, was among the many natural scientists who only much later realized that biology would never be the same again. Or perhaps it is more accurate to say that biology was invented on that day.

It was extinction that had put Darwin on the track of natural selection. In his early manuscripts, he doesn't hesitate to use terms such as "extirpation" and "annihilation." Yet, rather than dwelling on extinction, he almost seems to avoid it in his magnum opus, *On the Origin of Species*, published in November of 1859. For Darwin, extinction was inevitable, taken for granted. Natural selection would inexorably thrust some species aside: "As new species in the course of time are formed through natural selection, others will become rarer and rarer, and finally extinct. The forms which stand in closest competition with those undergoing modification and improvement, will naturally suffer most."[18]

When Darwin refers to such change, he is speaking in terms of geological time, of prehistoric eras that lasted millions of years, leaving their traces in the form of fossils. Major influences upon Darwin include his friend Charles Lyell (1797–1875), a Scottish geologist who opposed the idea of sudden and catastrophic change; Lyell maintained that the earth changed only very gradually.[19] While phenomena such as volcanic eruptions and earthquakes were sudden and dramatic, they were exceptional (at least in Britain, although certainly not in Iceland) and were themselves the culmination of long processes. Geological history and the story of life went hand in hand and moved slowly. Species had evolved and then died out, Darwin writes, in the distant past, long before the arrival of modern man:

> The species of the less vigorous groups, from their inferiority inherited from a common progenitor, tend to become extinct together, and to leave no modified offspring on the face

of the earth. But the utter extinction of a whole group of species may often be a very slow process, from the survival of a few descendants, lingering in protected and isolated situations.[20]

Darwin took little interest in the extinction of species, real or hypothetical, in his own time. For him, geological time and the unfolding of life-forms had little to do with bird-hunting off Iceland. Dinosaurs were one thing; great auks quite another.

For that reason, Darwin has less to say to us today than we might expect, as we foresee the extinction of many species, at the hands of humans, in the coming years. Darwin's conception of the world was static—as was remarked after his time—and not relevant to a world of rapid change generated by human activities.[21]

Extinction Observed

The great auk was one of the first species pushed "off the cliff" by humanity, its extinction observed by scientists more-or-less in real time. It is no surprise, then, that the great auk has come to stand for the concept of extinction in museums and in the public mind around the world, often with a heavy dose of loss and guilt: lost species remind us uneasily of humanity's predatory behavior—and of lessons that we may not yet have learned.

Arguably, the extinction of the great auk was inevitable, in view of massive European hunting of the birds for their meat, feathers, and oil in the 1700s and 1800s. By the time the international entourage of collectors became seriously interested in the species, its population was small and barely viable; collectors like Wolley and Newton did not kill off the great auk.[22] It has been suggested that environmental changes may have played a part—for instance, a drop in sea temperature affecting the bird's food sources.[23] Genetic research published by Jessica E. Thomas and her colleagues in 2019, however, provides no indication that any environmental factors played a crucial role. Their sequencing of DNA from great auks from all over their historical habitats points to considerable genetic diversity; only if genetic diversity had been low would it have been difficult for the species to adapt to environmental change. The authors conclude that hunting pressure by humans alone was sufficient to lead to the extinction of the great auk.[24]

When Wolley and Newton set off for Iceland in 1858, such an idea was unheard of. Species that were no longer seen where they were expected to be found were assumed to be hiding. When hunters in Europe, North America, and Greenland found the traditional breeding grounds of the great auk deserted, they assumed that the bird had simply gone elsewhere to breed that year. In the mid-nineteenth century, the consensus among ornithologists and amateur bird enthusiasts seems to have been that the great auks were hiding in Iceland. In 1854, for example, a paper was published in a Scottish journal under the heading, "The Great Auk Still Found in Iceland." The author, writing about great auk hunting off Iceland between 1813 and 1844, remarks that "there can be little question as to the great auk still existing in some numbers in Iceland; . . . we shall one day hear of some of our enterprising countrymen having overcome all difficulties, and returning home with a rich booty."[25] Discussions of this kind, about where species believed to be at risk of extinction may still be found hiding "in some numbers," have a familiar ring in our own times.

We now know that, while the great auk population suffered its greatest devastation at the hands of European hunters off Newfoundland in the early nineteenth century, the species fought its rear-guard action for survival off the coast of Iceland in the 1840s. Thanks to Wolley's *Gare-Fowl Books* and Newton's writings after the two returned from Iceland in 1858, the bird's losing battle was fought under the observant eyes of scientists.

Not that Wolley and Newton fully understood what they were chronicling. During their stay in Iceland, they pondered the meaning of "species," but "extinction" was not seriously on their minds. Not yet.

In 1855, John Wolley had visited Oslo, Norway (or Kristiania as it was called at the time), a town no less preoccupied with birds than Victorian Britain.[26] There, he met Johannes Japetus Steenstrup (1813–97), professor of zoology and one of Denmark's leading naturalists. Steenstrup had recently completed an important treatise on the great auk.[27] It was published later that year, throwing light on the bird's history on both sides of the Atlantic Ocean. Steenstrup had made his reputation by excavating historical garbage pits in which great auks featured. The origins of the great auk as a species and as a subject of study may be said to lie in Steenstrup's book. Wolley clearly listened to what Steenstrup told him about the great auk colonies in Iceland. As written in his soon-to-be-published book, Steenstrup believed that while "no dense colony exists any more; . . . the bird may possibly still live . . . off the west of Iceland, but that colony must surely be very small."[28]

As early as 1838, an article in a Danish journal—"The Great Auk in Iceland?"—had maintained that these once-common seabirds were "likely to become obliterated [*udslettede*]."[29] Significantly, while this reference to "obliteration" represents one of the earliest warnings about the rapid decline of the great auk in Iceland, the Danish word *udslettede* projected an image of erasing or flattening out, not quite what we think of as extinction. Again, the birds could have just gone elsewhere to breed.

For a while, several English terms signifying the end of a species were in circulation, including "extinction," "extermination," and "extirpation"; Newton used them interchangeably at times, as did many others during the Victorian period. As his work on the fate of the great auk progressed, however, Newton would become the "chief proponent" of the term "extinction."[30] Since then, "extinction" has become petrified in scientific discourse, as a fossil is in rock strata. No other term seems to offer a serious challenge. Search for "extinction" on the web today, and the Google search engine will quickly locate five billion hits. As far as can be ascertained, the definition used is the one that Newton established: Despite its long and nuanced history, the term "extinction" seems to have been colonized by concerns over loss of habitat as a result of human activities. Indeed, it is not far off to speak of "Newtonian extinction."

It is important, Newton insisted, to distinguish the prehistoric extinctions established by Anning and Cuvier from human-caused extinctions. Newton's breakthrough, in the wake of the Iceland expedition, was to move beyond the then-current notion of extinction as being the slow, long-durée consequence of natural forces—the notion highlighted in the works of Charles Darwin. Extinction, Newton reasoned, was not confined to deep time, to geological history or forces of nature, in the conventional meaning of the words. Extinction implicated humans, making them both complicit and responsible. Drawing upon the case of the great auk, Newton worked hard to publicize contemporary environmental problems. He helped to put bird protection on the political agenda in Britain and elsewhere. He was a founding member of the British Ornithologists' Union, one of the world's oldest and most respected organizations of its kind, in 1858, and of its journal, *Ibis*, the following year.[31] He wrote extensively about laws and other protective measures to avoid bird decline and habitat destruction, emphasizing both individual and professional responsibility.

Newton's work, in these ways, presages some of the ideas currently associated with the Anthropocene. That recent term—still contested but nevertheless rapidly catching on in both public and academic discussions—describes the geological epoch of our time, when human activities have written themselves into the geological record.[32] Often identified as starting at the beginning of plantation society, sometimes at the advent of industrialization or the dawn of the atomic age, the Anthropocene demands new perspectives and new thinking about extinction and the interdependency of species.[33]

Birds are relatively well-studied and monitored, and as a result, they are a useful barometer indicating the state of the planet. Among the many changes noted, bird populations around the world have dropped dramatically in the past half-century: in North America, their numbers have fallen by one-third.[34] The latest State of the World's Birds report (2022) shows that about half of the planet's bird species are in decline, and one in eight are threatened with extinction.[35] If we continue on in this way, the voices of birds will fall silent before long, both the beautiful melodies of swans and songbirds and the harsh squawks of seabirds, like the living relatives of the extinct great auk.[36]

Endlings

For many years after he returned from his Iceland expedition, Alfred Newton clung to the hope of someone, somewhere, seeing a great auk alive. But at last, he had to accept that neither he, nor anyone else, would ever see one again. In 1865, he wrote, still somewhat hesitantly, that the great auk should be seen as belonging to the past.[37] Among the documents inserted into the

Gare-Fowl Books is a copy of a letter acquired by Newton, written in Denmark in 1873. The writer says that she has inquired about drawings and documents relating to the great auk, and met with many influential people: "I met the Governor of Iceland . . . and asked him if there was no hope that Garefowl still might dwell within his dominion, but he said not the faintest hope was left: They are gone—extinct."[38] Was this declaration from the Icelandic authorities the equivalent of a death certificate—regarding the breeding population in Iceland at least?

In the 1990s, American physician Robert Webster launched the notion of an "endling," defined as the last person, animal, or other individual in a lineage. The idea came up because of patients who were dying and thought of themselves as the last of their family line. "Endling" seems to have stuck, outlasting competitors such as "ender," "terminarch," "lastoline," and "relict," and has seeped into popular culture.[39] When Alfred Newton's friend Charles Kingsley wrote his popular children's novel, *The Water-Babies* (1910), he did not know the word "endling." But he managed to capture its dual notion in presenting the death of the last great auk and the termination of a family line, whether a population or a species. The main character of his book, a great auk, says: "And I am the last of my family. . . . Once we were a great nation, and spread over all the Northern Isles. But men shot us so, and knocked us on the head, and took our eggs. . . . And soon I shall be gone, my little dear, and nobody will miss me."[40]

Endlings, almost by definition, drop out of sight; they are relegated to cabinets of curiosities or to the geological record. In the Darwinian perspective, they represent mere moments in the deep history of life—natural, to be expected, nothing to do with mere humans. But in the *Gare-Fowl Books*, Wolley and Newton captured the biography of two endlings—the last two great auks known to have been caught. Newton's intellectual career, after Wolley's death, would be guided by their story. Whereas Darwin sought to draw attention to variations within and among species to shed light on slow transitions from one life-form to another in the continuous process of adaptation, Newton focused on the recent histories of groups of species (he sometimes spoke of "zoological regions"), their decline and eventual disappearance—spelling out how catastrophe may be avoided. In this, he presaged a key concern of current extinction studies within the humanities.[41]

Yet, until recently, Newton's work has been strangely silenced and undervalued.[42] As American historian of science Henry M. Cowles has argued, by expounding the idea of two kinds of extinction—one natural, the other due to human impact—Newton presented the possibility that declines in nature might be reversed and at-risk species saved.[43] In such measures, Newton thought, experts in the natural sciences would surely take the lead. Influenced by his own fruitless hunt for great auks in Iceland, Newton introduced the idea that extinction is not a single event but an ongoing process—one that can be interrupted.[44]

For these reasons, it is vital for us to attend to the historic journey of John Wolley and Alfred Newton to the Reykjanes peninsula in the southwest of Iceland in 1858. Their dialogues with their Icelandic hosts about the fate of the great auk are more pertinent than ever, revealing both the blunders of the past and the clear danger to the future. The real weight of their quest rests on something far more fundamental than simply learning the fate of a pair of large, flightless birds that produced a single, beautifully patterned egg per year. By chronicling the disappearance of the great auk from its breeding grounds off the coast of Iceland, Wolley and Newton were elucidating the perturbed relations of humans and the rest of the animal world at a time of impending mass extinction.

CHAPTER 2

A VERY VICTORIAN QUEST

Before his untimely death in 1859, John Wolley did acquire a great auk egg. Cream-colored and ornamented with a unique pattern of swirls and spots in subtle shades of black and tan and brown, Wolley's egg came from Eldey—the very island off Iceland's Reykjanes peninsula that he and Alfred Newton had desperately hoped to reach the summer before. Wolley paid twenty-eight shillings for his great auk egg, about $900 at present-day values. Catalogued as number 4832 in the *Ootheca Wolleyana: An Illustrated Catalogue of the Collection of Birds' Eggs Begun by John Wolley*, the egg was collected during an expedition in 1835 (see plate 3). Carl F. Siemsen, a Reykjavík merchant, had bought the egg from an Icelandic hunter and sold it on to J.G.W. Brandt, a German collector. Brandt in turn sold it to a certain John Gould, who sold it to D. Barclay Bevan, from whom Wolley purchased it. Prized eggs like this one traveled widely, and a precise record of provenance had to be kept, just as in the case of works of fine art. Interestingly, though, unlike pieces of art obtained in former colonies, eggs do not ever seem to have been reclaimed and returned.

Alfred Newton, too, had a great auk egg to handle and study—and more than one. On Christmas Day 1861, he had a lucky break: he made a chance discovery of ten great auk eggs, uncatalogued, in a collection at Cambridge University, where he lived and worked. He was astounded. He immediately wrote to his brother Edward with the news:

> Only fancy a discovery I made the other day; it quite took away my breath! Going to Surgeons' Hall to inspect [Richard] Owen's dissection of a Great Bustard, I found [Thomas Henry] Huxley there, who asked me what I wanted. . . . Ascending to the topmost gallery of the innermost room, a glass case with birds' eggs met my eye. After looking at one or two grimy Ostrich's . . . I saw, as I thought, a nice model of a great auk, next to it was a prickly hen's, and then, on, on, on, as far as the eye could reach, great auks! To cut it short, there were *ten*, nearly all in excellent preservation, though one or two are a little broken. . . . As soon as my first emotions by the way were over I called out over the railing to Huxley and told him what I had discovered. . . . How they came here I don't know, but expect to make out; no doubt they are [from] Iceland. I always was sure of more being in England than I could trace.[1]

None of these eggs came into Newton's possession, but eventually he obtained seven great auk eggs of his own—an exceptional collection.

Newton, who studied great auk eggs with the devotion of a nesting mother bird, must have realized the importance of each egg's unique markings. They varied enormously, as if each egg had its own barcode. Natural selection came into play, it was later deduced: The individual patterns on each egg enabled the parent birds to distinguish their own egg and incubate it, ensuring the survival of the chick inside. Such artistic eggs also drew the attention of the people who made their way to the skerries where the birds nested; once collectors around the world heard about these ornately patterned eggs, they vied to acquire them. It is no surprise that illustrators and photographers have extensively portrayed the variety of great auk eggs.

The Allure of Eggs

Egg-collecting was one of the characteristic activities of the Victorian age, which commenced in 1837 when Queen Victoria ascended to the British throne. The British Empire was then in the ascendant, extending its tentacles around the globe; empires do not come much larger or more powerful. At the same time, the Industrial Revolution, while leading to the concentration and exploitation of workers in sooty factories, also united far-flung regions of the world by means of steamships, railways, telecommunications, and science, as well as military power. The era was shaped, as well, by the journeys made by inquisitive travelers who collected objects, classified and documented them, and sought to understand the diversity they exemplified. Samples from all over the world flooded into Britain, into public collections and private homes. The whole world was developing into a cabinet of curiosities, and birds— at least, their skins and eggs—had a central place in this new, global order.

Never have so many eggs been collected, before or since. People visited nesting colonies in droves, took samples of eggs from nests, carefully blew out the contents (killing the animal-to-be), classified and registered the eggs, and gathered them into private collections or public museums. Some egg-collectors were motivated by scientific ardor and research protocols, while others saw potential profit; some simply enjoyed being outside and exploring nature, and others sought out eggs on aesthetic principles. Some eggs are certainly more beautiful than others, and their diversity is captivating.

No doubt, the human fascination with birds relates to birds' eggs. Eggs are vessels containing embryonic life on the cusp of animal and not-animal. As a result, acquiring or losing an egg may carry deep symbolic meaning. Birds often nest in nearly inaccessible places, and collectors often break the fragile eggs they are carrying. The risk and adventure, however, offset the wastefulness in their minds. Debates on egg-collecting are thus more complex and imperative than disputes on bird-skins and taxidermied creatures—relics of animals that are no longer alive. In time, egg-collecting has come to be regarded as close to criminal. *Why would anyone collect eggs?* people have begun to ask, sometimes stealing the collectors' eggs in protest.[2] Today, taking birds' eggs for private collections is prohibited in many places.

In the Victorian age, however, egg-collecting was a popular family hobby, uniting children and adults (mostly boys and men). Collectors formed societies, and papers and periodicals were published. Some collectors were feckless amateurs, others punctilious and precise professionals. Yet by the mid-nineteenth century, something was brewing—a growing awareness, at least, of a rapid decline in supply and possibly a sense of impending disappearance of certain species, such as the great auk. An article by W.R.P. Bourne chronicles rapid changes in the price of great auk specimens, both their skins and their eggs, and establishes the average price of an egg sold during the 1830s.[3] Two decades later, when Wolley and Newton began to plan their journey to Iceland, the average price of a great auk egg was twenty-six times higher—compared to six times higher for the bird's skin.

Such rarity added to the egg's allure. Accounts of nature observation and egg-collecting are often recounted in dramatic terms of heroic achievement: They detail the special talents of the collectors; the harsh conditions experienced in the field; the battle with the elements in remote regions, on hazardous rocks, or surrounded by wild beasts. In Victorian times, educated young gentlemen of means were expected to travel to new territories (preferably solo), climb mountains, seek out strange and rare animals, find themselves in the beauty of the world, and return home having learned from their experiences. Some of these gentlemen travelers were both scholars and artists: writers and painters were inspired by the sublime, the wide-open spaces of wilderness, and many felt a need to record and classify what they saw. Many scientific studies grew out of such a Grand Tour, including those of John Wolley and Alfred Newton.

John Wolley (1823–59): "The Lord of the Eggs"

John Wolley was a classic example of the Victorian gentleman (see plate 4). His grandfather was Richard Arkwright, renowned for inventions that helped to modernize textile production; he has been called the father of the modern industrial factory system. John's father, also named John, had a keen interest in local history, and was a well-known collector of manuscripts. The young Wolley received, as a matter of course, a good classical education. He was sent to Eton, the world-famous English public school, where many leading members of British society—scholars, princes, and politicians—have been educated.

He entered Eton at age thirteen, a year before Queen Victoria came to the throne, and did well in his studies there, but his interests lay outside the classroom. From boyhood, he had been fascinated by the environment. During his six years at Eton, he was constantly wandering the nearby countryside, according to his own account: he was familiar with all the plants in the area, he collected insects and eggs, and he gathered all sorts of information.[4] Objects of natural history that he and his two elder brothers collected were preserved in a "museum"—as the earlier *Wunderkammer*, or cabinet of curiosities, was by then usually named—in the Wolley family home in Matlock, Derbyshire.

Between terms at Eton, Wolley traveled in Germany and Switzerland, gathering knowledge of birds and their eggs— and climbed Mont Blanc, a rare feat at that time. Those who knew Wolley praised him for his intelligence and daring. He was fearless, unhesitatingly scaling steep cliffs, and keeping snakes in drawers at Eton; he was said to swim like a fish in the sea and lakes. From his youth, all indications were that he would achieve great things in natural science—if he lived long enough to do so.

Leaving Eton in 1842, he went on to the universities of Cambridge, London, and Edinburgh, where he studied a range of subjects, as was common at the time, including law and medicine. It was at this time, in his early twenties, that he started to explore the fate of the dodo, a tall flightless bird that had been slaughtered en masse in the seventeenth century on the island of Mauritius in the Indian Ocean.[5] This striking, stubby-winged bird was killed and eaten, or captured alive for export, and before long it had died out. Little was known of the bird's appearance or behavior in Wolley's lifetime; it had vanished two hundred years before human-caused extinction was seen as credible or deserving of special discussion and scrutiny. But the dodo took on a new life in popular culture in the West soon after Wolley's death in 1859, featuring, for example, in Lewis Carroll's *Alice in Wonderland* (1865).

It may have been his study of the dodo that led Wolley to the great auk and to his focus on birds in general. For, although he passed his university examinations in medicine, he decided to put the profession aside. Instead, in 1851 he set off on an

FIGURE 2.1. John Wolley. (Wikimedia Commons.)

egg-collecting tour. He traveled alone for months in the Sami communities of Norway, Sweden, Finland, and Russia, collecting eggs and birds and making notes on his observations of the fauna of the far north. He understood the situation of the Sami, who migrated with their reindeer herds across national borders, violating the orders of the various countries, some of which were in conflict with each other. The reindeer, regardless of national borders, had to adapt to seasonal changes in their different grazing areas; the Sami had to follow the reindeer.

Wolley wrote many detailed letters from the high north to two aunts in England. He also corresponded with other naturalists, including Alfred Newton, and while his letters from the far north present a rose-tinted view of the birdlife there, Wolley does not conceal the problems encountered by egg-collectors. Wolley was an outsider among the Sami, and perhaps his zeal for collecting attracted unforeseen controversy and criticism. In a letter to Newton dated July 1853, probably written when he was with the Sami community in Muonioniska, Finland, near the Swedish border, Wolley wrote:

> Mosquitos, fleas, bugs, midges, dirty house, no bedclothes, no bread, sour milk, reindeer flesh raw and as hard as a board, are not luxuries. If you want to wash they bring you the same little sour bowl out of which you drink. All these little things make a bird-nesting expedition here very different from one when one leaves a comfortable English house in the morning to return to it in the evening.[6]

Egg-collecting is a less enjoyable pursuit, he warned his friend, when living in conditions very unlike what one is used to. The working environment was challenging for gentlemen who hailed from Eton, Cambridge, and Edinburgh, men from families whose riches were founded on industrialization and slavery.

Despite the discomfort of his quest, Wolley indeed returned from his sojourn among the Sami with what the author of the 1854 Scottish journal article, "The Great Auk Still Found in Iceland," would have termed "a rich booty": his luggage was laden with eggs and bird-skins, his notebooks full of detailed observations. He had a knack for getting to know the local people wherever he went: he talked to as many as possible, learning about people's lives, and questioning them about birds and eggs, hunting and collecting. Wolley contributed extensively to the northern communities he visited, sharing his ornithological practices and knowledge with both his assistants and about seven hundred local contacts.[7] During a difficult period, he organized fundraising for the support of the communities that hosted him.

The locals were impressed by Wolley's extensive knowledge and passionate egg-collecting; they called him "Lord of the Eggs," *Munaäija* or *Äggherren* (in Finnish and Swedish, respectively). For his expanding egg collection, it was important to preserve everything. Although it seems unethical to us today, Wolley sometimes took more than one egg and one bird-skin of the same species. He was not alone in this practice. Collectors dealt in rare eggs and birds, often selling part of their haul in order to subsidize their travel costs. Wolley had to pay his assistants a good wage for searching out birds and eggs for him, and in the Sami regions he sometimes employed dozens of assistants.

According to several Swedish and Finnish sources, during his stay in northern Sweden and Finland, Wolley had twin boys with a woman named Yliniemi from Muonio, Finland, who worked for his host.[8] Apparently, there is no reference to this in the limited sources available on Wolley in English. Wolley was said to have sent the mother a substantial amount of money before he died, as provision for his sons, who later emigrated to America. These sources were convincing, but they did not provide solid evidence such as a census or records of baptism and communion. As a result, the identities and fates of mother and sons were unclear. I wondered what happened to Wolley's sons?

The National Archives of Finland and the genealogical service MyHeritage provided information on births, baptism, and emigration at the time in question for the small town of Muonio, but there was no evidence of twins. On the other hand, a likely candidate was listed, born on 19 September 1857, a few weeks after Wolley returned to England. He was named Isak Liisanpoika Yliniemi (1857–1932) and was the son of a widow, Liisa Greta Henriksdottir Yliniemi. Records of baptism state that his father was "unknown." Isak, it turned out, set off for America in 1880, settling as a farmer in Becker, Minnesota. He married a Finnish woman, and they had thirteen children. It seemed that Wolley's son was found; everything seemed to match the narrative (except the story of twins). When contacted, however, Isak Yliniemi's descendants in the United States emphatically asserted that he was not the son of Wolley but of a married Finnish man from her home community. DNA tests indicated that the Yliniemis in America had Finnish origins, not British.[9] Wolley's affair, then, if there was such a thing, remains a mystery. His papers do not seem to mention it. Perhaps Wolley and Liisa Greta had become friends and he sympathized with the poor single mother, sending her money. Perhaps it was convenient to point to Wolley as the likely father; soon he would leave for good.

Before leaving the north, Wolley engraved his name on a historic stone using medieval runic script, as if he wanted to testify for eternity to his presence, his work among the Sami, and his attachment to the local community. The skillful runic inscription, which continues to receive attention, includes the following (in Swedish and English): "We are in [King] Oskar's land, liberated from Victoria of England. . . . This holy rock has taught Jon Volli of Matlok to craft runes." The reference to Victoria was probably a geopolitical observation, written during the Crimean War.

When Wolley carefully packed up his eggs, the collection weighed about a ton and filled an entire railway carriage.

Together with the later publication of *Ootheca Wolleyana*, Wolley's eggs have been deemed to be of outstanding cultural value.[10] His collection is almost unparalleled. Most of its more than ten thousand eggs come from the Nordic countries and the British Isles, but some were acquired from as far away as Siberia and North Africa. When the collection was being moved into its present premises in the Museum of Zoology, in the heart of Cambridge, England, some of the eggs were damaged in transit. Faced with a doorway too narrow for the storage drawers, the movers turned some of the drawers on edge; the eggs were thrown together, and broke. Each egg is worth a fortune, so the damage was enormous. Restoring damaged eggs is not easy.[11]

It was Alfred Newton who arranged for the surplus objects from Wolley's collection—rare items from Lapland—to be sold at Stevens's Auction Rooms in London, where collectors from all over the British Isles gathered once a year to make purchases and exchange views. Collectors vied to acquire eggs and skins of recently discovered species, and some fetched high prices. Rumors spread of profiteering, and ornithologists were criticized for taking part in the killing of rare birds which they themselves believed should be conserved. Often, they retorted that such stories were exaggerations.

For Wolley, Iceland was a dreamland of birds, where he would be able to see fast-flying falcons, enigmatic ravens, and gentle swans, and perhaps even the great auk, a bird for which he had no apt adjective, as yet. As soon as he returned from the lands of the Sami, Wolley set his sights on Iceland. When the opportunity arose, he started to learn Icelandic, probably at the urging of Newton, who had studied Scandinavian languages during his first years at university; he was sure that it would prove useful eventually.

FIGURE 2.2. Stevens's Auction Rooms, London, 1934.
(Photo: Errol Fuller.)

Alfred Newton (1829–1907)

Alfred Newton was the fifth son of a wealthy couple, William
and Elizabeth Newton, who had a total of ten children. In 1828,
the couple traveled in their private carriage from their home in
England to Pisa in Italy, holidaying in the manner of the upper
classes of the time, accompanied by servants and nursemaids.
On the way back from Italy via Switzerland, they made a stop
in Geneva, where Alfred was born.[12] They returned to England
the following year, and for a time, William Newton was a conservative member of Parliament for Ipswich in East Anglia. The
Newton family owed its wealth to slavery on the sugar plantations they owned on the Danish-ruled island of St. Croix in the
West Indies (now part of the US Virgin Islands). Iceland was

also ruled by Denmark at the time: the Danish empire crafted a strange nexus between north and south, the North Atlantic and the Caribbean, relating Iceland and St. Croix through mutual dependency.[13]

Newton studied at Magdalene College, Cambridge, from 1848 to 1852. (There is no indication that he was related in any way to physicist Isaac Newton, one of the leading lights of Cambridge University.) At Magdalene, Alfred lived in two rooms in a centuries-old building (see plate 5) that still houses the Pepys Library, a priceless collection of books and manuscripts acquired by diarist Samuel Pepys (1633–1703). While Magdalene College has a long history, it was small by comparison with the grand old colleges of Cambridge such as King's, Trinity, and St. John's. In Newton's time, students at Magdalene numbered about fifty, while the total population of Cambridge was just under thirty thousand.[14] Ambitious students strove to win scholarships at the major colleges in order to ensure themselves financial security and future prospects. Newton had to settle for a smaller and less illustrious college. Despite his family's wealth, he was, according to college records, a "pensioner"—an ordinary undergraduate who had to pay fees.

It was at Magdalene that Newton started to correspond with well-known ornithologists and naturalists of the time. A year after he graduated, he was awarded the Drury Travelling Fellowship, which enabled him to travel the world for the next ten years—to the West Indies, to Scandinavia, and with Wolley to Iceland, among other places, always on a quest for birds and their eggs. He retained his place at the college throughout his travels, and after 1866, when he was appointed to the first professorship in zoology and comparative anatomy at Cambridge University, he became a full member of the Magdalene faculty. While not highly paid, the post guaranteed him a living and a

FIGURE 2.3. Alfred Newton, 1857. (Courtesy of Balfour
and Newton Libraries, Cambridge University.)

home, and a suitable environment in which to pursue his research and writing.

His biography, *Life of Alfred Newton*, written soon after his
death by A.F.R. Wollaston, one of his former students at Cambridge, is an invaluable source. Despite its chaotic form (it
sometimes reads like a stack of letters), it details Newton's intellectual development, his academic connections, and his professional life, including the Icelandic expedition. A self-appointed

spokesman for the free birds of the air, Newton surely benefited from the slave economy, as the son of a slave-holding family. When his father, William Newton, died in 1862, his assets were valued at six thousand pounds, according to records of British slave-holders.[15] This was a considerable sum: sufficient to buy four hundred horses, or eleven hundred cows. But did Alfred Newton explicitly defend slavery and racism, or did he avoid facing the uncomfortable facts of life—focusing instead on birds? Perhaps his attitude was similar to that of his Magdalene friend Charles Kingsley (1819–75), who helped him acquire museum specimens from the West Indies. Kingsley was the author of *The Water-Babies*, published serially in 1862–63.[16] This children's fable, in which a great auk tells the story of its species, which is at risk of extinction, was hugely popular for many decades, before ultimately falling out of favor due to the ethnic slurs included.

The year before Alfred Newton traveled to Iceland, he visited St. Croix. But his attention was caught primarily by the birdlife—not by his patrimonial plantations or sugar production, let alone the entanglements of slavery in the West Indies. On St. Croix, Newton studied the behavior of hummingbirds, especially the green-throated carib (*Eulampis chlorolæmus*). One day, he watched a hummingbird that "flew into the room where I was sitting, and, after fluttering for some minutes against the ceiling, came in contact with a deserted spider's web, in which it got entangled and remained suspended and perfectly helpless for more than a minute, when by a violent effort it freed itself."[17]

Like the hummingbird, Newton had his own mobility problems. He had injured a knee at the age of five, romping with his brother in the library of their home at Elveden in Suffolk. Although the damage was initially regarded as minor, the injured leg was stunted, so Newton grew up lame, and walked with a stick. Does his own disability help explain his avid interest in the flightless great auk?

The Original Penguin

The two naturalists had seen relics of great auks; they had handled stuffed birds and dry bones. Still, they wondered, what does the living bird look like? They were keen to create their own image of the great auk, its appearance and behavior, and how it was perceived by its human neighbors on Iceland's Reykjanes peninsula. Would a visit to the breeding grounds fulfill their dreams? Planning the journey to Iceland, Wolley and Newton pored over papers and books containing images of great auks.

They must have studied the black-and-white drawing by Ole Worm, the Danish polymath who had sometimes walked his pet great auk on a leash. So far as they knew, Worm's was the only depiction ever made of a living great auk.[18] Worm could have been in no doubt about the bird's behavior—among humans in seventeenth-century Copenhagen, at least—though he left us no description of it. The oldest painting of a great auk in color, so far as is known, is a watercolor painted in the 1660s (see plate 6).[19] It is highly unlikely that Wolley and Newton ever saw it, as it was not widely exhibited. Its subject was the live great auk kept in the court of the Sun King, Louis XIV of France (who reigned 1643–1715), at Versailles.

Another powerful image, probably unknown to the larger world until the twentieth century, is that of Greenlandic artist Aron of Kangeq (1822–69). He knew the bird by personal experience; he had hunted great auks off southwest Greenland, near Nuuk. One of his well-known pictures, made after the great auk had vanished from Greenland's seas, depicts a legendary hunter hauling a great auk aboard his kayak in rough seas (see plate 7). In the Inuit legend, hitting the bird with his bird dart in a raging storm, the hunter lost his paddle, but getting hold of the bird, he regained his balance.[20] The painting may have been a wistful ode to the great auk, named in Greenlandic Inuit *esarokit-sok* ("Little Wing").

The discovery in 1991 of avian life-forms in Paleolithic cave art at Grotte Cosquer near Marseille in Southern France prompted archaeologists working there to talk about great auks (*grands pinguins*)—indeed the resemblance of the images to *Pinguinus impennis* seemed obvious. Visitors were stunned to see the striking images, as much as twenty-seven thousand years old, on the cave walls of a winding tunnel. Drawn on a shiny wall with charcoal, some of the images present the outlines of tall birds spreading out their tiny wings, as if they are ready for a fight or about to take a dive. These are probably the earliest images of great auks, drawn by people who had seen them alive, thousands of years before Ole Worm sketched his pet bird in Copenhagen.[21] Sooner or later, with rising sea levels, all these images will be destroyed, but now they can be observed and studied in a virtual museum nearby, a gripping replica of the real caves.

Archaeologists Anne and Michael Eastham have warned, quite rightly, that whatever the intentions of the artists in prehistoric time, "we have no reason to suppose that the characteristics represented in the picture were chosen to signify a certain species, as defined by modern zoologists. Paleolithic categories had no cause to follow our Linnaean rules."[22] Indeed, ethnozoology, the comparative study of the classifications of animal forms, testifies that a variety of systems have been developed through the ages and across cultures, following various kinds of logic and for various purposes.[23]

Artistic depictions aside, Wolley and Newton knew that the great auk was a large, flightless bird, standing about 33 inches tall and weighing about 11 pounds. To early observers, it seemed as if the bird's wings had been clipped, as if the flight feathers had been cut or removed. The stubby wings lay close to the bird's flanks, and apparently hardly moved when it was at rest; when it was in motion on land, the wings were flapped a little to help it maintain its balance. When swimming, the great auk was fast and agile; it moved like its much smaller relative the guillemot (*Cepphus grylle*).

By the sixteenth century, the great auk was known as *penguyn, pengouin,* or *penguin,* in Dutch, French, and English, respectively—possibly a reference to fat (*pingus*) and without wings or flight feathers (*in-penna*), but maybe from the Welsh *pen* ("head") and *gwyn* ("white"), a reference to the characteristic white spot between the bird's eye and beak. When European travelers in the southern oceans came across large, flightless birds that resembled great auks, they called them "penguins," and the name stuck. The great auk truly is, in other words, the original penguin.[24]

Pinguinus impennis became the bird's scientific name only in 1791, however; at first, the species was named *Mergus Americanus,* "American Duck," and later, *Alca impennis*—which is how Wolley and Newton referred to it, along with the common name "gare-fowl," a cognate of the Icelandic *geirfugl*. For about a hundred years in the eighteenth and nineteenth centuries, American whalers called it neither "penguin" nor "gare-fowl," but had a very different name for the bird.[25] That name, "woggin," appears in ten variant spellings in the logbooks of New England sea captains (see plate 8). Woggin was used both for great auks on the shores of New England and for penguins in South America. The mariners presumably first came across the great auk in northerly climes, then wrongly assumed that the penguins they saw in the southern hemisphere, based on their size and flight-lessness and harsh call (not to mention their wholesome and plentiful meat), were the same species.

The original range of the great auk was indeed quite large—though it did not span hemispheres. When, in Wolley and Newton's time, archaeologists unearthed relics of large, short-winged birds from kitchen middens at many different sites around the North Atlantic, they saw that the birds' bones were similar. The scientists concluded that the birds hunted on all these sites in Europe and North America had been the same—of a single species—although not at all related to the penguins of the southern hemisphere.[26]

Slaughterhouses

Archaeological research on bird bones in North America, the British Isles, Greenland, Denmark, and Iceland also revealed that the hunting of the great auk has a very long history.[27] Its breeding grounds were not chosen by chance. For thousands of years, offshore skerries and islands provided a refuge where the birds were largely safe from predators—although both Neanderthals and *Homo sapiens* hunted great auk in some numbers, for their meat and plumage, and for ritual use.[28] Some graves were covered in hundreds of great auk beaks, indicating that the bird was of high status. In the creation myths of the Penobscot people, indigenous to the east coast of North America, the great auk was a symbol of power and chieftainship. Anthropologist Frank Speck (1881–1950) wrote: "The great auk . . . is still remembered among the Penobscot as one of the legendary birds."[29]

The oldest maps of the North Atlantic, made in 1505 and 1520, show a Bird Island off the northwest coast of Newfoundland.

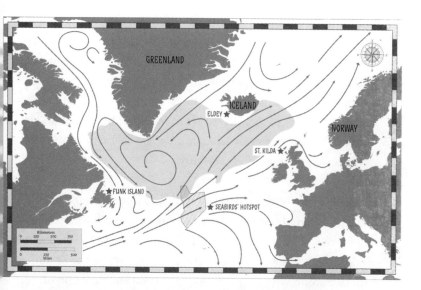

FIGURE 2.4. The North Atlantic. Great auk habitats and ocean currents. Three of the stars indicate the last known breeding grounds, at Funk Island, Newfoundland; Eldey, Iceland; and St. Kilda, Scotland. The "Seabirds' Hotspot" represents an area where many seabirds temporarily assemble on their migrations. (Courtesy of Matthías Ægisson.)

That eponymous bird was most likely the great auk. As Europeans increasingly sailed to the New World, in the wake of colonial expansion and improvements in ship technology, the nesting colony of great auks on this island—once the largest colony in the world—became vulnerable. On the island, the big, defenseless birds were herded like sheep into pens, gang-planked into boats, and clubbed to death. The meat was used for human consumption, and also as fish-bait, and the fat was rendered down for its oil, while the silky feathers were useful for stuffing quilts and pillows. French explorer Jacques Cartier, who sailed to New France in 1534, described a brief visit to Funk Island, near Bird Island, "Funk" being a reference to layers of stinking guano:

Our two longboats were sent off to the island to procure some of the birds, whose numbers are so great as to be incredible. . . . Some of these birds are as large as geese, being black and white with a beak like a crow's. . . . And these birds are so fat that it is marvelous. . . . Our two longboats were laden with them as with stones in less than half an hour. Of these, each of our ships salted four or five casks, not counting those we were able to eat fresh.[30]

Some of the birds were burned alive to loosen their valuable feathers. When the seafarers had filled their vessels with feathers and barrels of salted great auk meat, they sailed away. The meat was enough to feed the entire crew all the way across the Atlantic back to Europe, with some left over. The great auk fueled, in other words, European pillaging of the New World, feeding the very colonial system that would massively reduce the great auk population. Yet large-scale hunting of the birds for food in the sixteenth century does not appear to have played a crucial role in the demise of the Newfoundland great auks. In 1713, after two centuries of hunting by Europeans, the bird colonies were still going strong.

In due course, from around 1770, feathers of various wild bird species became a sought-after item in Europe for feather beds, and for the next decade hunting expeditions were made every year to Funk Island and its vicinity. By 1785, warnings were heard that, unless this feather-hunting was prevented, the bird colonies of Newfoundland would collapse, "for this is now the only island they have left to breed upon."[31] At its peak, the breeding population of great auks around Newfoundland is believed to have been about 100,000 pairs per year. By around 1800, the great auk had disappeared from the region.

The next wholesale destruction of great auks occurred at Geirfuglasker, the Great Auk Skerry, off Iceland's southwest coast. In the summer of 1808, a group of men led by an "English Viking" went to the skerry and slaughtered all the birds they could catch. Another expedition five years later was triggered by geopolitics. Hostilities between England and Denmark had culminated in the Battle of Copenhagen in 1801, the primary source of conflict being competition among slave-owning sugar barons for control of the high seas. The odds were against the Danes and, as a result, there was a severe shortage of food in both Denmark and the Faroe Islands, which depended on Danish agricultural exports. Desperate as a result of the war-time lockdown, and with his people reduced to near starvation, in 1813 the governor of the Faroes sent a vessel to Iceland (which was also ruled by Denmark at the time). The crew of the schooner *Færöe* landed at the Great Auk Skerry and found an abundance of great auks. They killed all they could and loaded the boat full, leaving many dead birds on the rocks. Due to a storm, the boat shifted course and sailed north along the Reykjanes peninsula to stop briefly in the capital city of Reykjavík, before continuing home to the Faroes.

Alfred Newton labeled this hunting trip a "massacre"; the boat had "on board . . . no less than twenty-four Gare-fowls, beside others which were already salted down."[32] Awareness of this event may have triggered his notion of an impending unnatural extinction: if he and Wolley did not reach Iceland soon, they might miss their chance of ever seeing a great auk or collecting its egg.

CHAPTER 3

AN AWKWARD TRIP TO ICELAND

John Wolley and Alfred Newton were more familiar than the average person with the natural history of the great auk, although certain aspects of it remained mysterious. Will we have the good fortune, they wondered as they looked forward to their Icelandic expedition, to get close to at least one great auk—to examine it from an ornithological viewpoint, make accurate drawings, smell the bird's odor, observe its behavior—and perhaps acquire an egg or a bird, whether live or stuffed? They realized that their only hope of seeing the birds for themselves was to visit Iceland during the breeding season. They decided to aim for Eldey island, a tall rock with vertical cliffs off Iceland's southwest coast, in the same archipelago as the Great Auk Skerry on which the birds had been massacred in 1813. By the 1840s, Eldey was the great auk's best-known nesting grounds; why the bird no longer favored its eponymous skerry was, to the two British naturalists, at the time unknown.

The two explorers had acquired a copy of a map of Iceland made in 1844, based on surveys by Björn Gunnlaugsson (1788–1876), an Icelandic mathematician and surveyor trained in

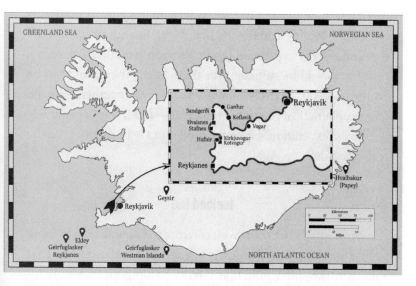

FIGURE 3.1. Iceland. Great auk nesting places and the key sites of the British great auk expedition of 1858. (Courtesy of Matthías Ægisson.)

Denmark. This map shows Eldey and many other islands and skerries off the Reykjanes peninsula, as well as the nearby West-man Islands. It is a fine work of cartography, the result of many years' work. Such a high-quality map, printed in color, attracted visitors; those who were planning a visit to Iceland could feel that they were already halfway there. Newton found the map "beautiful."[1]

He and Wolley considered whether they should ride on horseback across the highland interior to the north of Iceland, in between their ornithological tasks, or whether they should travel around the coast. Would they be at risk in the uplands of encountering outlaws, even the mythical hidden people (*huldufólk*), admired and feared by many Icelanders? Would the glacial rivers of the southeast prove an obstacle? What kind of equipment would they need: tents, sleeping bags, reindeer pelts?[2] They decided to take as little luggage from Britain as possible. Wolley said that he had about a hundred and twenty pounds for the expedition (about $20,000 in modern currency), which should be sufficient for them both. They could not be sure, however, of always being able to travel together. It would be problematic if they were separated—but it was "a genuinely *awkward* expedition,"[3] quipped Wolley in a letter, punning on the bird's name.

Iceland Ho!

In early April 1858, the travelers wrote to Copenhagen to book their passage on a steamer headed for Iceland later that month. The ship would take twenty passengers, they were informed; they were to meet it in Scotland. The departure was delayed for a few days, but that only made it seem more convenient: They would not have to stay as long in Reykjavík, Wolley remarked, with nothing to do but study Icelandic while they waited for the great auks to nest on Eldey at the end of May.[4] Few vessels sailed to Iceland in the mid-nineteenth century, and this was their only option.

They met in Edinburgh on April 20, then sailed the following day from Leith harbor aboard the steamer *Victor Emanuel*. It was a one-deck iron cargo and passenger vessel, driven by both steam and wind. Built in Scotland in 1856, it was then on its first postal trip serving two of the satellites of the Danish state, the Faroe Islands and Iceland. The morning of 21 April 1858 was unusually warm. As the ship steamed out of the mirror-smooth Firth of Forth, Wolley and Newton passed the time by observing the familiar seabirds, particularly the velvet ducks (*Melanitta fusca*).

By coincidence, another passenger on the ship was the German polymath Konrad Maurer (1823–1902), sometimes called the

FIGURE 3.2. *Victor Emanuel.* Danish vessel (later renamed *Arcturus*) on which John Wolley and Alfred Newton traveled from Scotland to Iceland. (Courtesy of National Museum of Iceland.)

father of Nordic studies. Maurer would go on to write a journal of his travels in Iceland, which would lie unpublished for over a century. The three travelers had much in common and became well acquainted on their journey. Wolley and Maurer were the same age, thirty-five years old. Maurer wrote in his journal:

> four Englishmen had come aboard, ornithologist Wolley from London and his companion Newton; also two merchants, one of whom was part-owner of the ship. They were all pleasant and cheerful people, especially Mr. Wolley. . . . I later had the opportunity to get to know him and his companion somewhat better, and that acquaintanceship is one of my best memories from my journey.[5]

The days grew longer. For some time, there was nothing to see but the wide ocean, until the Faroe Islands rose above the horizon. When the ship arrived at Tórshavn, the principal town, many passengers went ashore—along with twenty letters from Denmark, the first mail to reach the islands that year. Wolley and Newton were enchanted by the narrow streets, the low turf-roofed houses, and the Faroese in their fine clothes wielding pitchforks and spades as they prepared their gardens for the summer. In the evening, they dined with Sheriff Hans Christopher Müller, whom Wolley had met during a previous stay in the islands.

From the Faroes, the ship set a course for Iceland. The sea grew rougher, and the vessel was tossed about on the waves. Wolley and Newton were cheered to spot a bird they had never seen before, the little auk (*Alle alle*), a dainty relative of the great auk, with the power of flight. The vessel steamed past the Westman Islands in poor visibility, and the travelers had no chance to see that archipelago—which included yet another skerry named Great Auk Skerry.

Passing Eldey, the Meal Sack

Continuing west, the ship swerved around Eldey, known to foreign visitors as the Meal Sack. Newton wrote in a long letter to his brother Edward:

> About noon we passed the celebrated Meal Sack, but we must have been nearly two miles from it. It is certainly well named, for in one direction it has very much the sort of look of a sack half filled, with the sides turned down. . . . On the landward side runs out a low shelf or rock where the great auk is supposed to have bred.[6]

There were few passengers on deck. Many had disembarked in the Faroes, Newton reported, while others were seasick: "A considerable number of Icelanders, both male and female" kept to their berths. "Wolley, however, behaved remarkably well and was never fairly under the weather."[7]

This route, the Reykjanes Race (*Röst*), is no joke to sail in rough weather, even aboard a large ship. Powerful currents from different directions converge here, and mere humans have no answer for the stormy waves. Open rowing boats were often lost with all hands here during the fishing season, a devastating blow to the families and communities of the crew. But Wolley and Newton stayed out on deck to watch the mystical Eldey pass by in the ocean spray. Heavy seas or no, Wolley observed the island with his characteristic precision, as the vessel "came with the rock bearing N.W. by W.1/2W"—he must have had a compass at hand. "I made a rough sketch," Wolley continues in his journal, "which is now before me. A few minutes previously A. Newton had made a sketch of it on the other side of the same piece of paper: *Sketch: Meal Sack (or Eldey)*. . . . From this point of view the rock was seen to be full of ledges and bold faces in which were overhanging projections, and the dung of sea birds whitened it in many places."[8]

The tall island abounded with birds (see plate 9). Fulmars (*Fulmaris glacialis*) and razorbills (*Alca torda*) squawked in polyphonic chorus, though sometimes drowned out by the noise of the ship's steam engine and the pounding waves. Gannets (*Morus bassanus*) flew back and forth, apparently without purpose, while black-legged kittiwakes (*Rissa tridactyla*), eiders (*Somateria mollissima*), and Atlantic puffins (*Fratercula arctica*) were also seen, the puffins restlessly flying low over the surface of the sea, as if they had difficulty maintaining flight. Like the little auk that Wolley and Newton had spotted in the Faroes, the puffin, too, is a relative of the great auk, though much smaller, with dainty little wings.

The two ornithologists must have been delighted by the sight of Eldey, the fabled breeding ground of the great auk, which they purposed to visit in a few weeks. They observed the

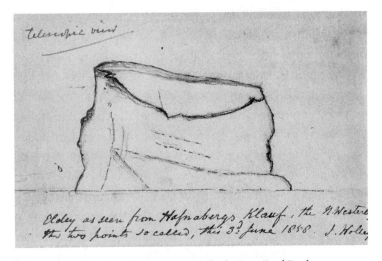

FIGURE 3.3. Sketch of Eldey. John Wolley's *Gare-Fowl Books*. (Reproduced by kind permission of the Syndics of Cambridge University Library.)

surrounding waters with care, just in case a large, stubby-winged bird might appear in their path; but they saw none. The birds that swam around the ship appeared to have no interest in it— nor any fear.

Some of the passengers laughed when a young girl climbed briskly up on deck, and pointed to a farmstead on the shore. Though she was pale and tired after the difficult journey, her face shone as she exclaimed: "Ah, there is Ness! There I was born!"[9] There is no record of her name. Following her gaze, Wolley and Newton observed the scattered farms and communities of the deeply indented Reykjanes coastline. They must have noticed the rough cliffs—and the absence of any safe harbors or landing places. Probably they were also struck by the lack of trees, the flatness of the terrain, and the tiny houses below impressive mountains in the distance. Equipped with their map and compass, they sought to imagine what would happen next. It was from this stretch of coast in Iceland's far southwest that the latest expeditions had been made to hunt the great auk—they would soon make their way there.

In Reykjavík

The *Victor Emanuel* anchored in the Reykjavík harbor, and the passengers were taken ashore in a small boat. A crowd of towns-people had gathered—it was not every day that a steam-powered oceangoing vessel arrived—leaving Wolley and Newton ill at ease; they were carrying a considerable sum of money. They expected to pay a whole boat crew to venture into hazardous waters, quite apart from all their other expenses. Although they were dressed for horseback travel in all weathers, they were obviously gentlemen, wearing long overcoats and formal head-gear, and carrying the leather briefcases of scholars. The two British naturalists were aware that they clearly stood out.

For accommodations, they were told to go to The Club, otherwise known as the Scandinavian Hotel, a two-story timber-framed building with a hipped roof. There they were offered two small rooms; one was shared by Wolley and New-ton, while Konrad Maurer occupied the other. The guests were comfortable in their rooms, but at dinner they were dismayed to be offered eider (*Somateria mollissima*) and red-breasted merganser (*Mergus merganser*) to eat. They were not aware—and nor was anyone else—that great auk bones lay underground just a stone's throw away: nearly a century later, archaeologists unearthed the bones from an old kitchen midden on a building site in the city's center.

At the time of Wolley and Newton's visit, Reykjavík was be-ginning to develop from a small town into a capital city.[10] About 1,500 people lived there, or approximately two percent of

FIGURE 3.4. Reykjavík harbor, 1889–94. (Photo: Sigfús Eymundsson. Courtesy of Reykjavík Museum of Photography.)

Iceland's population. Most of the buildings were one- or two-story tarred wooden structures. Many had decorative curtains at the windows. Above the town rose a windmill, a sign of growing industrialization. The Alþingi, Iceland's ancient parliament, abolished by the Danish authorities at the end of the eighteenth century, had been reinstated in 1845, and convened each summer in Reykjavík. The Learned, or Latin, School, whose roots lay in a cathedral school founded in the Middle Ages, had also been transferred to Reykjavík. The old bishoprics of north and south Iceland had been abolished at the end of the eighteenth century, and a new cathedral was built in Reykjavík, where the new Lutheran bishop of Iceland presided.

The establishment of these three key institutions was a sign of Iceland's progress toward self-determination and ultimately independence, after centuries of Danish rule. "New trends,"

FIGURE 3.5. An early daguerreotype of Reykjavík, around 1881. (Photo: Sigfús Eymundsson. Courtesy of Reykjavík Museum of Photography.)

wrote Bishop Jón Helgason, "have arrived here with changing times. The capital has become far more strongly of our nation than before—a truly *Icelandic* town. Yet all the leading merchants in the town are still men of foreign origin."[11] Denmark had dominated commerce in Iceland over the centuries, with some exceptions (Britain and Germany also played a role at times); this only began to change with the granting of Free Trade in 1855.

Despite the town's progress, visitors were not impressed. British Reverend Sabine Baring-Gould, who traveled to Iceland in 1862, comments:

Reykjavík is a jumble of wooden shanties, pitched down wherever the builder listed. . . . The moment that the main thoroughfares are quitted, the stench emitted from the smaller houses becomes insupportable. Decayed fish, offal, filth of every description, is tossed anywhere for the rain to wash away, or for passers-by to trample into the ground.[12]

A growing number of the townspeople were relatively prosperous, but others lived on the brink of poverty. Drunkards were seen wandering the streets, accosting foreign tourists and fingering their fine clothes.[13]

But the conditions were certainly no worse than those Wolley had borne among the Sami, and he and Newton promptly ventured out to meet locals who could give them useful advice about their quest. On their second day in town, they called upon the principal of the Learned School for coffee, after which they went to meet the Danish governor, Count Jørgen Ditlev Trampe. Both men gave them a warm welcome, conversing in a mix of English and Danish.

The year of Wolley and Newton's visit was marked by harsh weather, with a cold and stormy spring, and for the next three days it poured with rain. The travelers needed to acquire a range of equipment and supplies, as well as arranging safe transport for the eggs and birds they expected to acquire. But it was impossible to reach their ship, still at anchor in the bay, in order to fetch their luggage. The principal of the school, taking pity on them, invited them home for a "bit of bread," which turned out to be a lavish dinner—food and drink, claret and punch—in "terrible quantity," as Newton put it. Four more leaders of Reykjavík society had also been invited to the dinner, which did not end until two o'clock in the morning.

The following day, they called on a pharmacist who was said to have an interest in birds. He had several bird-skins, including a gyrfalcon (*Falco rusticolus*) and a harlequin duck (*Histrionicus*

histrionicus), but his visitors were not overwhelmed. On his wall, the pharmacist had a picture of the latest great auk that had been caught; they were told that the painting had been made in 1844 or 1846 by a French artist named Vivien. The fate of this historic picture is unknown; it may not have been a very attractive work of art. Wolley and Newton felt that the depiction of the bird was poor: The eyes were too big, and the wings too long. The dead bird had clearly been held in place to approximate a natural pose for the painting to be made. They seem to have felt that the bird had not been treated with sufficient dignity. Perhaps their critique of the work of the French artist reflected a growing tension in Victorian thought. Ideas of taste and representation were in a state of flux, paving the way for a new sense of observation, science, and objectivity.[14]

Making Arrangements

Within a few days, the two British naturalists had made the acquaintance of most of the town's dignitaries. As a rule, they got straight to the point, asking about the rare great auk. The pharmacist informed them that in recent years two expeditions had been made to Eldey: The men had succeeded in landing on the rocky ledge on the island but had not come across any great auks. The Danish governor told them that a fixed price had been paid for each great auk expedition, plus a sum for each bird caught. He himself had seen three of the birds since being appointed governor eight years before. These birds, caught in the Westman Islands, had been the property of a Reykjavík merchant.[15] This sounded promising, if it proved correct.

The governor warned them, however, against sailing out to Eldey themselves, as it was too hazardous. Having crossed the Reykjanes Race on their way to Iceland, they did not doubt his

word. Venturing across in a rowing boat would be quite different from their experience aboard the steamer. What were they
to do? Who could ferry them to their destination?

Luckily for them, Vilhjálmur Hákonarson of Kirkjuvogur
was in town. Perhaps this was not pure coincidence. German
zoologist Frederik Faber, who visited Iceland and wrote a well-
known account of his trip, stated that in 1820 he had negotiated
with Vilhjámur's father, Hákon, renting a boat for an excursion
to the great auk sites close to the Reykjanes peninsula.[16] When
arranging the trip, Wolley and Newton had contacted Grímur
Thomsen, an Icelandic poet and literary scholar. Thomsen sent
a "friendly" letter of recommendation to Ketill Jónsson at
Kirkjuvogur, who happened to be Vilhjámur's foster-father.
Jónsson, in turn, must have alerted Vilhjálmur to the matter of
assisting the visitors.[17] Vilhjálmur, then, went to Reykjavík
partly to meet with the British visitors, launching their historic
collaboration.

Sometimes known as "the Wealthy," Vilhjálmur Hákonarson
(1812–71) made his fortune from the sea. In his log of his fishing
expeditions, he wrote that he had first rowed out at the age of
sixteen in a six-oared boat, during the winter fishing season
of 1829. After that, he became a foreman, in charge of a boat and
a crew. At the time of writing, he states that during the recently
ended winter fishing season of 1858, he had rowed out forty-
eight times, every other day on average. "I still have the same
boat, and I plan to keep with it," he says. She was a lucky vessel.
He had rowed her to Eldey several times. Hákonarson was said
to be "decisive in disposition, yet level-headed and calm; firm
when challenged, and unwilling to back down when he sees
that right is on his side; utterly reliable in business and his word,
and the most honorable of men."[18] He was no newcomer to
great auk hunting. On his first expedition, he had caught about

FIGURE 3.6. Vilhjálmur Hákonarson. National Museum
of Iceland. (Drawing said to be the work of Sigurður
Guðmundsson, around 1865. Courtesy of National Museum
of Iceland.)

twenty birds himself. He would become a good host and an
excellent "witness," to use Wolley's term.

Hákonarson told Wolley and Newton about "his latest trip
when he caught a great auk." They listened with care and arranged for him to house them at Kirkjuvogur and take them out
to Eldey. On May 2, Newton wrote to his brother:

We have . . . seen a man who lives at Kirkjuvogur, the nearest
village to the Meal Sack. He has made four trips there, the

last in 1856, when no Geir Fugles [gare-fowls] were found: On the previous occasions, 24, 7, and 2 were obtained. We are to go to him in about a fortnight, and then to make the expedition in two boats. I do not think there is much risk if, as we shall do, we wait for favorable weather.[19]

The visitors were fortunate enough to be offered an escort to Kirkjuvogur. Dr. Jón Hjaltalín (1807–82), Iceland's director of public health, volunteered to ride with them: he was heading to the region to visit leprosy patients, the disease being widespread in Iceland at the time.

The Collectors' Contract

One place that Wolley and Newton were keen to investigate was another island named Great Auk Skerry, this one off Iceland's southeastern coast. In Iceland, the number of reefs with the same name are an indication that the great auk was seen in some way as special. Farmers hunted seabirds and collected eggs in huge numbers from these skerries; sometimes the bird hunters made three expeditions in the same summer. But the Great Auk Skerry that Wolley and Newton wanted to visit—better known as Hvalbakur (Whale's Back)—was in the East Fjords, the full length of the country away from Reykjavík and the Reykjanes peninsula, from which they hoped to reach Eldey. Wolley and Newton could devise no way of getting to the east of Iceland as well as to Eldey, given their schedule.

A possible solution arose: Eiríkur Magnússon (1833–1913), a theology student at the clerical seminary in Reykjavík, had grown up in the East Fjords, near Hvalbakur, and might be sent there on their behalf. The seminary's director arranged to let Magnússon off from his spring examinations, and the two

British naturalists commissioned him to travel by horseback to the east, talk to people in the region who knew about great auk hunting there, and go out to the skerry. Magnússon undertook to collect eggs and to catch and preserve birds in accordance with the visitors' instructions. He was to be paid three hundred Danish rixdollars to cover his expenses and remuneration.

Magnússon's contract with Wolley and Newton, which takes up four pages in the *Gare-Fowl Books*, is surprisingly difficult to parse, despite the detail. It seems to have been jotted down in a hurry, with several words crossed out and overwritten. It says, in part:

> Hr. M. will endeavour to ~~obtain~~ browse as many eggs as possible of this bird; and skins, and whole bodies according to the agreement already made verbally, that is to say, if only one bird is ~~to be~~ found it is to be preserved whole . . . , if two birds are found the second is to be carefully skinned, prepared with essential soap, and its body placed in . . . preserving mixture. . . . The third bird is to be preserved whole as the first, the fourth bird to be skinned as the second and so on alternately to the seventh. If more birds are obtained one of them ~~at least~~ should be saved alive, and the others skinned. If ~~there are more than~~ after eight birds are obtained more still ~~remain~~ are left on the rocks, it is wished that on no account more than half of the ~~surviving~~ survivors be killed— and in any case it is particularly desired that no more birds be destroyed than can be well preserved in skins or in the mixture.
>
> Every egg and every bird of the Geirfugl is to be wholly property of Mr. Newton and Mr. Wolley, and Hr. Magnússon undertakes on no account to let them be otherwise disposed of. Mr. Newton and Mr. Wolley hope that Hr. M. . . . . will

make careful observations on the Geirfugl ~~skerries~~ rocks, and also the birds themselves according to the memorandum which they have given him; that he will place the result of such observations in detail upon return in the Danish language, and leave them at ~~Mr. N. and Mr. W.~~ the disposal of Mr. N. and Mr. W., to be used for the benefit of science as shall seem best to them.[20]

This remarkable historical document—unmistakably drafted by Newton, judging from the handwriting, signed and dated by the three men in Reykjavík, and later inserted into the third volume of Wolley's notebooks—is the only contract of its kind known from the two collectors. The complicated instructions for Magnússon are simultaneously illuminating and puzzling. Were Wolley and Newton simply overexcited by the opportunity to explore another one of the fabled breeding grounds of the great auk, without taking the trouble themselves to travel there? Or did they truly expect Magnússon to discover such a healthy breeding colony that he could kill eight great auks and take an additional half of the "survivors" on the rocks, plus collect "as many eggs as possible," without damaging the species' prospects? Where, now, are the two naturalists' fears that the great auk was in decline, rare, or even vanishing? Did their contract adequately reflect the biblical spirit of the game laws expressed in the book of *Deuteronomy*—"if you happen to find a bird's nest, do not take the mother with the young"—which would inform Newton's later work on bird protection?[21]

Given the excessive demand in the market for great auk specimens, it is understandable that Wolley and Newton were keen to nail down the issue of property rights. As they most likely had heard, on Hákonarson's "latest trip when he caught a great auk," the hunter had made only a verbal agreement with the

Reykjavík merchant Carl F. Siemsen; returning with two birds, he had then sold them to someone else, who was conveniently closer than Reykjavík and offered a higher price. Did Wolley and Newton envisage selling off their "rich booty" of eight great auks and "as many eggs as possible" at Stevens's Auction Rooms in London, as a means of recouping the expense of their Icelandic venture?

Clearly, by specifying "on no account more than half of the ~~surviving~~ survivors be killed—and in any case it is particularly desired that no more birds be destroyed than can be well preserved," Wolley and Newton wanted to maximize their catch without ruining the colony. They were not warranting a massacre. But aiming for eight birds, or more, seems a bit excessive, given recent reports on the scarcity of the species. Strikingly, Newton never mentions this contract in his later accounts of the expedition—despite the fact that, until the end of his life, he remained in near-daily contact with Eiríkur Magnússon. In 1866, the Icelander moved to Cambridge and a few years later, with Newton's help, obtained the position of university librarian. Perhaps the two men found it embarrassing afterward, as Newton's work on the great auks unfolded and he became deeply involved in environmental activism.

"Unnecessary Hospitality"

While in Reykjavík, Wolley and Newton dove into the local library, inquired about documents and sources, and took notes on long reports in Icelandic and Danish about the hunting of the bird. They also engaged a teacher at the Learned School to give them lessons in Icelandic. He began by teaching them Icelandic bird names, explaining that the common metaphorical expression "to stick one's neck out," *að teygja fram álkuna,* literally means to act like an inquisitive auk. Wolley, remarked Newton in a letter to his brother, was progressing well:

> Wolley is studying his task, which is a saga and seems to be written intentionally for beginners, as it opens, "There was a man named Grim." The pronunciation is the most difficult thing I have ever heard, it beats Finnish into fits. . . . I want to know a few words as up the country there will be no one who can speak Danish.

While the men were keen to interview local people themselves as far as possible, they soon realized their language skills would not take them very far and decided to employ an interpreter. They were fortunate to hire Icelander Geir Zoëga (1830–1917), who by then was making a name for himself as a guide to foreign tourists; in addition to Danish, he also spoke English. Zoëga would go on to become a highly successful entrepreneur in tourism, retail, and the fishing industry.

Tired out after long days of Icelandic study, and much coffee and talk, Wolley and Newton would sit down together each evening to make notes of what they had learned. Three weeks in the little town of Reykjavík was a long time. "Of course," wrote Newton, "it was a great bore being weatherbound in the metropolis as it was not lively, and a very bad place ornithologically speaking. Besides this we were almost in a chronic state of intoxication from the unnecessary amount of hospitality we had to endure, but as it was all meant as civilly as possible one had nothing to do but abide it."[22]

Wolley appears to have found it easier to adapt to the local ways. Wrote Newton: "Wolley requires gentle managing, he is too fond of delaying things."[23] Here, as among the Sami, Wolley felt a need to get to know the people and their customs; he initiated conversations on subjects of mutual interest, and told

FIGURE 3.7. "English Party at Reykjavik." Drawing by Bayard Taylor, 1862. (Library of Congress, Washington, DC. Public domain.)

people about himself, thus establishing a relationship of trust and preparing the ground for further visits. He was working, in fact, like a modern anthropologist.

Their final task in Reykjavík was to select the horses for their field trip. They had postponed the decision, in order to avoid having to keep the horses at pasture or on fodder at their own expense. But the time had come. Nesting season was about to begin: hopefully, the great auk awaited. On May 19, Wolley and Newton set off on horseback for the southwest corner of Iceland, for a district on the Reykjanes Peninsula known as Suðurnes (literally "South Peninsula"). They were pleased to have the company of several local men to ease their way: not only did Dr. Jón Hjaltalín, the director of public health, and his assistant accompany them, but a veterinarian who was dealing with a severe outbreak of sheep scab also came along. Rounding out their party was their guide and interpreter, Geir Zoëga. The great auk quest had begun in earnest.

"A Less Perishable Inheritance"

John Wolley had begun writing his *Gare-Fowl Books* in Reykjavík, unaware of their eventual scale and number, yet hopeful for their historic significance—although neither he nor Alfred Newton could anticipate their crucial role in informing future discussions of extinction and environmental politics. They were not just tracking a rare—or missing—species but also identifying a key moment in the history of human-animal relations. In a summary article that he published shortly after the expedition, Newton made this bold statement:

> Whether the Gare-fowl be already extirpated or still existing in some unknown spot, it is clear that its extinction, if not already accomplished, must speedily follow on its rediscovery. I have therefore to beseech all who may be connected with the matter to do their utmost that such rediscovery should be turned to the best account. If in this point we neglect our opportunities, future naturalists will reproach us. The mere possession of a few skins or eggs, more or less, is nothing. Our science demands something else—that we shall transmit to posterity a less perishable inheritance.[24]

That "something else" was preserved in the pages of the *Gare-Fowl Books*, and transmitted via the conclusions Newton drew from his and Wolley's expedition.

While at first Newton's views did not gain much traction, ten years later he would deliver a lecture, "On the Zoological Aspect of Game Laws," that had a potent impact in Britain, leading to organized efforts to protect endangered species. In this monumental but brief address (it was only two pages long)—read on a Sunday morning, 22 August 1868, at a lively meeting of the British Association for the Advancement of Science (BAAS) in Norwich, England—Newton suggested that, while the notion of "Game Laws" commonly referred to regulations adopted "to control the relation of man to wild animals," the destruction of the animals concerned was rarely discussed. Not surprisingly, he drew attention to the fate of seabirds:

> With reference to sea-fowl, a certain amount of sentiment may be confessed. No animals are so cruelly persecuted. At the breeding-season they come to our shores, throwing aside their wary and suspicious habits, and sometimes settling far inland. No one has ever complained of them as injurious. . . . We thank God that we are not as Spaniards, gloating over the brutality of bull-fights, whilst we forget the agony inflicted on thousands of innocent birds on our coasts to which that of a dozen horses and bulls in a ring is as nothing: The enormous demand for the feathers of sea-fowl by the modern fashion of ladies' hat plumes has added to this cruel destruction.

Newton offered two kinds of remedies. The first, he argued, was to engage public opinion: "The public should feel that they have an interest in the protection of wild animals, especially during the season of reproduction. The decrease of these animals, however, is often attributed to secondary causes, and not to direct slaughter." His second recommendation was to legislate a "close time," a seasonal suspension of hunting, based on zoological research and professional advice:

> This plan has been adopted in several countries, . . . as shown by the Game Laws of Switzerland, Norway, the United States

of America, and several British colonies. . . . If the present state of things continues much longer, far greater changes will take place with regard to the fauna of this country than most persons suspect, and they will be changes for which the zoologists of future generations will not thank us.[25]

Drawing on what he had learned searching for great auk in Iceland, Newton forcefully argued that laws on "close time" during the breeding season were necessary to avoid extinctions. Apparently, those attending the Norwich meeting agreed. Newton's message is still remembered and honored, as "Alfred's battle," by British naturalists.[26]

The BAAS established a "Close-Time Committee," comprising Newton and five other scholars and environmentalists. This committee helped draft the "Sea Birds Preservation Bill," which was ratified by Parliament on 24 June 1869. While Newton was not alone in this effort, he took a leading role in fleshing out the legal details and the nuances of language, as well as in publicizing the rules and their implications for preservation.[27] The act specified that "[w]hereas the sea birds of the United Kingdom have of late years greatly decreased in number; it is expedient . . . to provide for their protection during the breeding season." The key paragraph (S-2) runs as follows:

Any person who shall kill, wound, or attempt to kill or wound, or take any sea bird, or use any boat, gun, net, or other engine or instrument for the purpose of killing, wounding, or taking any sea bird, or shall have in his control or possession any sea bird recently killed, wounded, or taken, between the first day of April and the first day of August in any year, shall, on conviction of any such offence before any justice or justices of the peace in England or Ireland, or before the sheriff or any justice or justices of the peace in Scotland, forfeit and pay for every such sea bird so killed, wounded, or taken, or so in his possession, such sum of money not exceeding one pound as to the said justices or sheriff shall seem meet, together with the costs of the conviction.[28]

Partly based on earlier legislation and foreign experience, the act was the first of its kind in Britain, specifically designed to reduce the effects of shooting and egg-collecting during the breeding season. It was welcomed by the public and, as a result, became a model for further legislation.

Moving from the protection of seabirds to that of wading and shore birds, though successful in the end, proved trickier; the preparation of a new bill along these lines in 1872 was met with strong resistance. At the time, the rules on hunting such birds were based on local conditions and principles.[29] Some birds were more privileged than others, reflecting a pecking order of classification based on social class and aesthetic principles. A powerful upper-class constituency regarded some of these birds as hunting quarries and their pursuit as a traditional right. For poor landless people, in contrast, hunting birds in this class was a matter of subsistence.

While in Newton's view the shore birds bill showed the moderation required in environmental diplomacy, "pretty well [hitting] the mean between extreme opinions," others have pointed out its aristocratic bias, in that it pitted working-class hunters against noble "sportsmen," landowners, and scientists (like Newton himself) who busily collected birds and eggs.[30] Newton was criticized for being hypocritical and protecting his own turf.[31] His double standard was indeed evident early on, in the striking contract on great auk hunting and egg-collecting in Iceland's East Fjords—during the breeding season—that he and John Wolley signed with Eiríkur Magnússon before they left Reykjavík in May 1858. As British ornithologist Tim Birkhead (b. 1950) observes, the elite were "fuelled [by] a ruthless, single-minded quest for samples": "Blinded by the light of their own fervour, it was almost as if Victorian ornithologists failed to see the paradox that, in the process of shedding light on birds, they were casting a dark shadow over their own history."[32]

CHAPTER 4

ARRIVING

Cambridge University Library keeps the *Gare-Fowl Books* with care—as is appropriate for old and irreplaceable manuscripts. The notebooks, now more than 160 years old, cannot be photocopied or scanned by library patrons. Yet, surprisingly, no printed copy exists of these vital historical documents. The notebooks are considered original literary works under UK law and cannot be published for the time being. Scholars can, however, acquire a digital copy from the library, on condition that only they have access to it; in that form, the *Gare-Fowl Books* run to about 900 photographs.

The Alfred Newton Archive, which includes Wolley's papers, is extensive: the catalogue extends over 174 densely spaced pages; the documents number in the thousands. The *Gare-Fowl Books*, collectively, are item number 2. The five notebooks vary in thickness and size, with covers of different colors. Two parts of the *Gare-Fowl Books*, the most extensive, comprise interviews with people Wolley and Newton met in Iceland and the documents they acquired. The other three parts are devoted to press coverage, auctions, letters, and so on, relating to the great auk. The five books, in varying formats, give the impression that the project grew beyond Wolley's initial expectations. In places, the paper is worn and the text smudged from being carried in the author's saddlebags while in Iceland, in all weathers.

Wolley's first book begins: "This book commenced in Reykjavík 30th April 1858 by me John Wolley just intended for notes on Alca impennis," *Alca* being the Latin genus name assigned the bird by Linnaeus; it was officially reassigned to its own genus, *Pinguinus* (though Wolley apparently disagreed), in 1791. Likewise, Wolley opted to use an old common name for the bird, "garefowl," instead of the more recent "great auk," introduced in 1768 by Welsh naturalist Thomas Pennant (1726–98)—perhaps because that is what his Icelandic informants called it.

No doubt Wolley intended to consult his notebooks in the future, to recall names and places, or details given by his informants. In aid of his future self, he occasionally made a note between the lines, cross-referencing similar material elsewhere in the *Gare-Fowl Books*. The books also functioned as a journal or diary, with regular entries. A diary offers an opportunity for time alone with one's thoughts at the end of the day; Wolley must have looked back over his experiences, writing down what he deemed too important to be forgotten.

Wolley's notebooks are almost all written in his own hand, and the interviews with Icelanders are largely his work. But occasionally, either in the field or later, when he reviewed the material after Wolley's tragic death in 1859, Newton added comments or explanatory notes. Wolley was sometimes in a hurry. On his travels about the Icelandic countryside, he wrote in faint pencil, which he later traced over in ink when he returned to his base at Kirkjuvogur on the Reykjanes peninsula. That is generally helpful, although the double writing is sometimes a problem for the present-day reader. Occasionally, Wolley interpolated pages from a smaller notebook, which appears to be otherwise lost. Wolley and Newton were like living photocopiers. They made copies of letters and old documents in several languages—Icelandic, Danish, French, and German—that they came across on their visits. They copied even old maps, historical documents, and drawings of great auks, with remarkable accuracy, all of which appear in the *Gare-Fowl Books*.

Unfortunately, Wolley's notebooks do not offer much insight into his personal reflections or observations in the field, nothing compared to the long and detailed letters he had sent from the Sami communities in Scandinavia to his family and friends in England. While Newton sent several letters to England, during their stay in Iceland, Wolley apparently did not; he was too preoccupied. As Newton notes in his memoir of Wolley:

> The country around possessed but few attractions for the ornithologist; but Wolley was indefatigable in seeking for information from the mouths of persons who had formerly visited the Skerries, and was successful in procuring from them many valuable and interesting particulars relating to this bird.[1]

Wolley, in Newton's words, was busy "seeking information at the fountain-head." Right after his opening words, "This book commenced," Wolley documents his discussion with the Danish captain on the way to Reykjavík. The captain said that "he had been something like 100 times round Cape Reykjanes but had never seen Geirfugl, but once seen 3 skins with a merchant at Keflavík, from which place boats had gone to the rocks."

Otherwise, the opening pages of the *Gare-Fowl Books* are cluttered with chaotic notes on hunters' memories of their hunting expeditions, with scribbled dates and catches of birds and eggs. They also record memories written in Icelandic by local priests, including "Tales about the Great Auk and Great Auk Skerries," listing fatal voyages in 1629 and 1639 during which several boats were lost and a number of great auk hunters drowned. This was not encouraging, as Wolley and Newton were counting on having the opportunity to land on Eldey themselves to observe the great auks breeding there.

Guests at Kirkjuvogur

Riding southwest from Reykjavík, Wolley and Newton progressed slowly. The spring was cold and rainy, and, although it was mid-May, the grass was still too sparse to provide enough grazing for the horses. The first day, they traveled less than thirty miles, on horses that were in poor condition. They stopped from time to time to water the horses, but the animals refused the brackish water. Perhaps that is one reason, they thought, that the Icelanders tended to opt for stronger drink.

The director of public health was a big man who required no fewer than three horses, stopping frequently to change his saddle to a fresh one, while Wolley had two and Newton one.[2] An accident delayed them, when the veterinarian fell from his horse, and the animal landed on top of him. On another occasion, they were held up when the director of public health had to detour to a farm to visit a leprosy case. The two British naturalists decided to accompany him. That evening, they were invited to dine at the farmhouse. When they found that they were being served birds, they asked to be shown the heads and discovered that they were eating small shore birds: purple sandpipers (*Calidris maritima*) and turnstones (*Arenaria interpres*).

Sometimes their route followed well-beaten tracks between farms, but often they had to pass through jagged lava fields, dark plates of rocks frozen in the middle of a volcanic eruption a long time ago (see plate 10). In some cases, they managed to skirt the rocks, taking a slightly longer but easier route through erratic moss or grass. The lava fields were difficult for the horses, generating deafening sounds. In heavy rain, the soft tracks were even worse; the horses seemed unsure of where their hooves were landing. Falling with the horse was no fun.

Along the way, there was much to catch the eye, such as the perfectly formed conical volcano named Keilir (The Cone) in the distance—a handy sign for wayfinding for riders as well as sailors, unless everything was surrounded by depressing fog. But overall, the landscape they rode through seemed to be limited to either a rugged, dark, stony terrain or various kinds of moss and grass in subtle shades of green, brown, and yellow. Sometimes the ever-present ocean added a roaring sound, reminding the visitors of their rough journey to Iceland. Did they find this picturesque and pleasant? Perhaps at times, but more likely the British travelers found the environment to be horrifying—with its complete absence of trees and flowers. Occasionally, they came across farmhouses covered with grass roofs, as if they were buried in the land, belonging to the planetary crust. Arriving at the first farmhouse on their way, they would have been stunned by the small entrance, the turf walls, and the dark and narrow subterranean channels between the spaces of people and their domestic animals: cows, sheep, and horses.

They would have liked to have had a camera, the latest in new technology: "Certainly a photograph would be the only thing to give an idea of the look of these places," commented Newton.[3] No doubt they were aware of *The Pencil of Nature* (1844) by William Henry Fox Talbot, a landmark publication in the history of photography—the first book to show pictures "drawn" by light alone. Wolley and Newton would have liked to fix the image of the great auk on paper by the wonder and apparent precision of photography, to prove to the world that the bird still existed and to establish how it "really" looked. But while early cameras were said to provide an unmatched "objective" account of the real world, potentially resolving complex and long-standing problems associated with biological description and the identification of species, they were large and unwieldy; it would have been impractical to try to transport one on horseback or in an open rowing boat. There were no roads through Reykjanes on which a wagon could travel.

They saw many birds along the way and invariably paused, however briefly, to take a better look. Once Newton shot a purple sandpiper and examined it carefully. For men such as Wolley and Newton, classically educated at elite British schools, the study of birds was a matter of course. Birds had a special place in the cultures of ancient Greece and Rome. Aristophanes's play *The Birds* depicts an outbreak of obsession with birds, or *ornithomania*: the characters strive to sing like birds, take the names of birds, and yearn to have wings of their own. The Greek word *ornis*, from which the word ornithology derives, can mean both *bird* and *augury*. Birds are portents and symbols for many aspects of human existence. In the excavated Roman ruins of Pompeii, murals have been discovered depicting eighty species of birds, as well as bird-themed ornamentation on various objects such as clay vessels, seals, sculptures, and mosaics. Birds were seen as emissaries between humans and the gods: "Winged words," as Homer put it, flew about, carrying messages of many kinds. Homer uses this image more than a hundred times, until it verges on being a cliché.[4] The image is also familiar in its opposite form: words that never gain wings, and so do not take flight.

Continuing their journey in search of the flightless great auk, Wolley and Newton rode past the fishing community of Keflavík (near the current site of Iceland's international airport) and called at the nearby church farm of Útskálar, where they met with the pastor, the Rev. Sigurður B. Sívertsen (1808–87). A pillar of the community, Sívertsen had taken the initiative in the

founding of an elementary school (schooling was not yet compulsory in Iceland). He was also a prolific writer, keeping the
annals of the surrounding district of Suðurnes, in which he recorded the weather and historic events, such as drownings at sea
and the loss of fishing boats in storms; he also compiled his own
memoirs. Wolley and Newton found him to be quite knowl-
edgeable about the history of bird-hunting in the area, as well as
the local topography. Others on the farm pointed out a refuse
heap where there were bird bones—probably including those of
great auks. Wolley and Newton set to work with their shovel, but
apparently found no great auk bones on this occasion.

After two days in the saddle, the travelers reached the manor
house at Kirkjuvogur (Church Inlet). Here, the British visitors
would stay, as Vilhjálmur Hákonarson's guests, from May 21 to
July 14, with the exception of two short intervals. Part of the
hamlet of Hafnir (Harbors)—which consisted of several houses
of farmers, tenants, and landless laborers, each house having its
own name—Kirkjuvogur was one of the major estates in the
district, comprising a large cluster of turf and wooden buildings.[5] Below the farm is a hollow where boats were beached
between fishing seasons.

The land was poor for farming, but the sea on the doorstep was
an endless source of food. According to the Icelandic *Register of
Estates* compiled in 1703, the grassfields of Kirkjuvogur "are at risk
of destruction due to drifting sand from all directions. Meadows
are none. Summer pasture near the farm of the poorest quality."
But, on the other hand, it was possible to row out to fish "all year
round, and the landing conditions are good. The boats of the
farmers here go out as is convenient to them, and sometimes in
the fishing season vessels from elsewhere sail for a fee."[6]

Travelers who are guided around the Icelandic countryside
by locals find landmarks at every step: hills, hollows, and rocks,

FIGURE 4.1. Kirkjuvogur in Hafnir on Reykjanes, the key site
of the Iceland expedition of 1858. Photo from 1873. (Photo: Sigfús
Eymundsson. Courtesy of National Museum of Iceland.)

all with their own placenames. But the small seaside settlement
of Kirkjuvogur was no backwater. Letters sent from the outside
world would get there, sooner or later. During Wolley's stay, for
instance, he received one that was forwarded by his friend in
the Faroes, Sheriff Müller. On the envelope was written:

> John Wolley, Esqr
> Hunting Geirfugl
> Reykjavik
> or Keblavig.[7]

By the spring of 1844, Hákonarson had taken over as foreman
of the Kirkjuvogur fishing boat and crew. He had learned from
his predecessor, Brandur Guðmundsson (1771–1845), how best
to sail out to Eldey and land on the island, and where to seek the
great auks. Guðmundsson was a man of multiple talents. He was
a fine boatwright, said to have built many of the local vessels. In
Icelandic, these boats were said to have been "begotten" by him
(*undan honum*), perhaps a sign of personal attachment to a

FIGURE 4.2. Letter addressed to John Wolley, "hunting for great auk, Reykjavík or Keblavíg"—that is, Keflavík. (Reproduced by kind permission of the Syndics of Cambridge University Library. Copyright Cambridge University Library. Photo: Gísli Pálsson.)

critical component of the household economy. A report that the Rev. Sigurður Sívertsen wrote (in Icelandic) for Wolley says of Guðmundsson: "Built one hundred and thirty-three boats. Died of *kvefsótt* [a bronchial infection]." Altogether, Guðmundsson led five expeditions to Eldey and caught more than forty great auks. Wolley quotes Hákonarson as saying, "Brandur said there was no need to look for great auks anywhere but on the big ledge under the rock."[8] The ledge is still visible, on the east side of Eldey.

"Odd Birds"

After Hákonarson had welcomed his guests and they had settled into their accommodation in the lofty manor house at Kirkjuvogur—a grand farmhouse by comparison with its neighbors, with window ornamentation and a dormer on the roof—they quickly made plans for sailing out to Eldey. "Our arrangements are completed," wrote Newton. "We are to have two 10-oar boats for greater security, and 16 men in each, so that some may rest; thus with ourselves and Zoëga there will be 35 souls embarked on the enterprise. With these precautions, I think the risk is reduced to a minimum."[9]

Wolley and Newton learned that the French surgeon and explorer Joseph Paul Gaimard (1793–1858) had acquired five great auk skins during his Icelandic expedition of 1838.[10] Would they be equally successful? They were well aware that much could go wrong despite Hákonarson's expertise. Wolley wrote from Iceland to his friend in the Faroes, Sheriff Müller, explaining their plans. Müller, a keen naturalist like Wolley, replied: "It gives me great satisfaction hearing you had got well settled in Iceland, and I wish to God you may return safe from your adventurous trip."[11]

The times were hard in Iceland. In the *Annals of Suðurnes* for 1857, the Rev. Sívertsen wrote: "The harshest of winters. Severe weather and lack of food. People here were in want, more than at any other time. . . . There were many storms, and in one of them a vessel was lost. . . . The boat was swept away on the current and foundered off the . . . headland, with eight men aboard." The year of Wolley and Newton's visit was not much better: "Very harsh winter with poor catches, though with some exceptions."[12]

Now, in mid-May, the oarsmen from Kirkjuvogur and the nearby farms were ready and waiting: They had prepared their oilskins, two 10-oared boats sat ready for action, and Hákonarson's wife, Þórunn Brynjólfsdóttir, had prepared enough food for them to remain at sea for a week. Their destination, Eldey, might prove decisive for the understanding of the great auk and, possibly, the general study of rare and disappearing species. All they needed now was good weather—which failed to come.

As they waited and waited for the weather to turn, it must have become clear to their Kirkjuvogur hosts that these visitors were "odd birds"—not the usual type of Victorian British tourist (see plate 11). Journeys to Iceland had been popular ever since the days of Sir Joseph Banks (1743–1820).[13] Banks, who had joined Captain Cook on his successful, historic expedition on the *Endeavour* to the South Seas from 1768 to 1771, was scheduled to travel with Cook on another ambitious voyage, trying to circumnavigate the globe as far south as possible, but due to a disagreement, Banks decided to head for Iceland instead to explore its volcanoes. Banks's visit in 1772 was the first expedition undertaken by foreign naturalists to the country. He and his entourage set off on the *Sir Lawrence* on 12 July 1772, the very same day that Cook sailed south.[14] Iceland had been placed on the map, as an interesting terrain for Western explorers and naturalists.

Banks became friends with many Icelanders, with their help collecting numerous botanical and geological samples, as well as manuscripts. One of the members of the expedition joked about the festive food they were offered one day, the flesh of whale and rotten shark: "This is either boiled, salted, or dried in the air, looks very much like rusty bacon, and has so disagreeable a taste, that the small quantity we took of it, drove us from the table long before our intention."[15] If Wolley and Newton had a similar experience, they did not mention it in their accounts. Likewise, Banks and his colleagues said little or nothing about Icelandic bird life. As Newton would remark in his essay "Notes on the Ornithology of Iceland," published in *Iceland: Its Scenes and Sagas* in 1863: "Though several British naturalists of no mean repute have visited Iceland, I believe that hitherto no connected account of its Ornithology has ever been published in the English language."[16]

Unlike their predecessors, Wolley and Newton were not much interested in botanizing or visiting volcanoes—they were in Iceland only for the birds. Newton, particularly, was single-minded and set in his ways, an introverted loner, with his pipe and tobacco always at hand. He had a neatly kept beard, and moved slowly, leaning on an ornamental walking stick, due to his lame leg. His hosts at Kirkjuvogur were probably astonished to learn that their distinguished visitor planned to embark on a dangerous sea journey in search of a bird that apparently had gone elsewhere! Though he certainly had many fine qualities, Newton was no doughty hero, to set out across stormy seas in an open boat; nor was Wolley, despite his daredevil ascent of Mont Blanc as a boy. Yet here they were, undismayed. At Kirkjuvogur, Wolley was his usual sociable, chatty, and restless self, striding across country in search of birds, with his rucksack and telescope, a notebook in his pocket.

In 1837, a pamphlet published in Copenhagen, printed in Icelandic, urged the need to teach Icelanders to swim: *Professor Nachtegall's Swimming Rules, augmented and adapted for Icelandic needs.* The translator of the text was the renowned Icelandic poet and naturalist Jónas Hallgrímsson, whose father had drowned when he was a boy. Hallgrímsson writes in the foreword:

> It is hardly an exaggeration to say that fourteen or fifteen years ago there were not more than six men in the entire country who could save themselves if they fell into a puddle where they could not touch bottom. . . . There were those who felt it would be more-or-less criminal to teach seafarers to swim, for that would only mean they suffered more if they fell into the sea. Nothing could save those who were destined to die, they said, nor could anything bring death to one who was destined to live.[17]

Wolley could swim like a fish. But what of Newton? The North Atlantic was not a cozy Victorian aquarium.

The Ocean as Aquarium

The great auk was an oceanic creature, spending most of its life in the open sea, as Wolley and Newton well knew; but they had little idea what that life was like. While mathematicians and astronomers had clearly established the movements of faraway heavenly bodies by the nineteenth century, the ocean remained a profound mystery. For ages, humans had viewed the ocean from its edges, looking out from the shore. During the colonial era, the human view of the ocean vastly expanded, thanks to European fleets and explorers crossing the seven seas, but in the Western imagination, the ocean itself remained two-dimensional, an immense seascape for travel, commerce, and communication. Humans had minimal interest in the watery underworld, a substantial part of the planet, mainly because of the difficulty of access.

The Victorian aquarium, a lively *Wunderkammer*, partly overcame that problem of access. The term itself, a contraction of "aquatic vivarium" (enclosures in which aquatic animals are kept), was coined by British naturalist Philip Henry Gosse (1810–88). Gosse, who created the first public aquarium at the London Zoo in 1853, is sometimes referred to as the David Attenborough of his day. His book, *The Aquarium: An Unveiling of the Wonders of the Sea* (1854), in which he described the construction and maintenance of aquaria in detail, caused an "aquarium craze" in England.

Gosse argued in the opening statement of his book: "The habits of animals will never be thoroughly known till they are observed in detail."[18] Before long, the emerging field of marine biology became accessible to both aspiring scientists and members of the general public. Tank keeping became a popular hobby, providing an opportunity for laypersons as well as professionals to collect, enjoy, and study a variety of aquatic forms as their lives unfolded in plain sight in ordinary living rooms, as well as in fancy museums. Gosse went on his own scientific Victorian quests, although he didn't travel far, exploring the coastlines and ponds of Britain to obtain samples of aquatic life. He was a skillful artist, illustrating his works with wondrous images of marine life forms.

In *The Victorian Aquarium* (2021), literary scholar Silvia Granata situates Gosse's *Aquarium* and its appeal in the social context of the era. Victorians, she argues, were not attracted to the aquarium—as we might be today—thanks to its relaxing or soothing effects; they wanted action and unmediated access to life in the tank. Artists and scholars often interacted directly with aquatic creatures, by putting their hands into the tanks. The aquarium "provided a surprisingly rich vista on mid-Victorian culture," an updated Noah's Ark domesticating the exotic ocean, bringing carefully selected fractions of it into plain sight and facilitating public engagement with the inhabitants of saltwater. This was genuine crowdsourcing, in modern jargon.[19] Granata points out that while Victorian students of the ocean drew upon existing discourses of animals, "marine species did not always fit into them (for instance, although *domestic*, they could neither be seen as proper *friends* nor conceptualized as *servants*)."[20] It became pertinent to speak of the birth of the aquarium as a general scientific model of the oceans—as a phenomenon to be subjected to systematic gaze and scrutiny.[21]

Perhaps the aquaria craze itself helped to foster the idea of the oceans as gigantic and vulnerable tanks in danger of destruction, through the overexploitation and extinction of aquatic life-forms. It may also have contributed the pathbreaking HMS *Challenger* expedition of 1872–76, which continues to inform environmental discoveries and concerns.[22] During their long voyage through the oceans, *Challenger* scientists and crew launched oceanography, exploring oceanic conditions and the landscape of the sea bottom, cataloguing more than 4,000 previously unknown marine species.

Virtual Aquaria

Wolley and Newton were aware of Gosse's work on aquaria. Newton and Gosse visited each other and corresponded about their travels and mutual interests, including birds and the Gospel. As Newton was preparing for his expedition to St. Croix in the Caribbean, Gosse sent him detailed guidelines on how to keep the tiny and fragile hummingbirds alive, an issue that had long bothered many ornithologists.[23] "As to the capture of hummingbirds," Gosse said, "you will find it the easiest thing in the world."[24]

Newton hoped to acquire a great auk for the London Zoo, before it was too late. A great auk wouldn't fit well with the fish in an aquarium tank, given its inclination to travel fast and far. Perhaps it was a marginal figure, a category mistake in the era's rigid classification of life-forms, a flightless avian figure inhabiting the oceans—somewhat like flying mammals, including the endangered flying foxes of Australia (Pteropids).[25] Academic circles in Europe regarded early reports of flightless birds as dubious, if not an impossibility. Faced with realistic drawings of flightless emus and cassowaries from the South Seas, the Paris Académie des Sciences concluded in 1807 that they were pivotal in silencing skepticism about "these extraordinary beings which seem to contradict our prior ideas."[26] Newton and Gosse

might, in fact, have thought the great auk to be more at home
with the birds in the aviaries, sometimes known as flight cages,
which also became popular in the Victorian age. Like aquaria,
aviaries—whether built in public museums or privately
owned—allowed people to gaze at, and learn about, the lives of
the world's creatures. Whether grounded or suspended, aviaries
also displayed their owners' power, status, and wealth.

What was the place of humans in this world of fish tanks
and flight cages? When they studied the great auk, did Wolley and
Newton and their colleagues see humans as part of the grand
three-dimensional tank of the global oceans, living alongside
the great auk, other seabirds, various fish, and Moby Dick, or
were we—are we—detached observers? Clearly, the ability to
understand the elements of the tank and to model the interactions and energy flows involved helps to set meaningful levels
of long-term resource use. Sometimes, however, such scientific
models "leak"—somewhat like the classic aquaria developed by
Gosse and his contemporaries, spattering parts of the contents
into living rooms and museums. For one thing, ecosystems in-
volve chaotic processes that are difficult to model, much less to
manage. The family aquarium may be easy to deal with, but it's
hardly the North Atlantic.

Although we still struggle with the fundamental questions
of where humans belong—inside the tank with all of earth's
creatures, or outside, looking in through the glass—Wolley and
Newton clearly saw themselves in the role of detached observers, endeavoring to understand hunting practices and ideas in
a foreign context, and destined to leave when they were done.
Like proper zoologists at the time, they were keen to identify
breeding colonies, migration routes, and catch totals. Yet they
were caught in webs of Victorian science and religion, which
biased their understanding, so far, of human responsibility for

the great auk's extinction. For them, the flightless great auk that traveled throughout the North Atlantic Ocean belonged inside the aquarium (or perhaps the aviary), while humans did not. They did not question the dualistic godly design, by which humans were fundamentally separated from the rest of the animal kingdom. Extinction, for Newton, despite his new insight based on his work on the great auk, was still dual and halfway, mostly "natural"—though he would soon begin to argue that humans had a responsibility to keep it that way. The human-caused mass extinctions of our current age were, in the mid-nineteenth century, simply unthinkable.

CHAPTER 5

REBOOTING

Few people land on Eldey nowadays, but some go to great lengths to see the island where the last known great auk was killed. In her book *The Sixth Extinction* (2014), American writer Elizabeth Kolbert discusses her arrangement with a fisherman in the town of Sandgerði, on the Reykjanes peninsula. Through a complex network of contacts, he agreed to take her to the island, "but only if the weather was fair; if it was rainy or windy, the trip would be too dangerous and nausea-inducing, and he wouldn't want to risk it."[1] His boat was small, but with a powerful motor and an extra life raft, a modern luxury. Once they arrived at Eldey, however, Kolbert gave up the hope of landing; the sea was beating against the rocks. She now realized that the white ripples on the surface of the island were gannets (*Morus bassanus*); the colony, one of the largest gannet colonies in the world, was made up of thousands of pairs. Atop the island was a webcam that had been set up to stream a live feed for bird-watchers. Wolley and Newton would have loved the opportunity to observe the bird colony from an office in Britain, intently searching for great auks. But the camera was not reliable, Kolbert was disappointed to learn. The birds, she was told, did not

FIGURE 5.1. Eldey island. The great auk nesting place, the "underland," is to the right. A webcam is visible on the top to the left. (Photo: Sindri Gíslason.)

like the camera, so they would "fly over it and shit on it," blocking the lens.[2]

Several years after Kolbert's visit, in December 2022, a group of five natural scientists and environmental observers landed with a helicopter on Eldey. The tiny island hadn't been visited recently due to the Covid pandemic, but this time the visitors were shocked by what they saw.[3] The top of the island was littered with colorful plastic and dead birds. During recent breeding seasons, the gannets had increasingly constructed their nests using orange and green fragments of fishing gear that they discovered in the ocean nearby. Practically every nest was made of plastic, and hundreds of birds had died on the spot as they got entangled in the web they had constructed (bird flu may also have contributed to the deaths). If this continues, the striking white cover of the island will fade away. The old label of the "Meal Sack" will be a misnomer.

Stuck Ashore

At Kirkjuvogur in 1858, the two British naturalists were on edge, always aware that they might set off for Eldey at a moment's notice. Wolley and Newton differed somewhat in their views: "Wolley is much more sanguine about success than I am, and I think more than he has a right to be," Newton wrote to his brother, "but at the same time I am not more desponding than I have always been about it."[4]

One morning, the crew prepared to launch the boats; they were about to push off when Hákonarson, the foreman, suddenly called off the trip. After all their preparations and the excitement of finally readying the two boats, this must have been a shock. But the foreman's instincts proved correct: before long the sea grew too rough. The other local boats remained on shore as well: rough seas precluded any fishing. For the fishers, the unpredictable downtime between trips, waiting for the right conditions, was a well-known hiatus of uncertainty: a period of lethargy—or levity. It even had its own Icelandic term, *landlega,* or "stuck ashore." The masters of farms and fishing vessels complained that their workers were ungovernable in such circumstances, until they could go back to sea.

But the winter fishing season ended without conditions clearing; spring arrived. Some of the workers were supposed to travel north for the hay-making season, but they hung around to see if Hákonarson would head out to Eldey in search of birds. Not a single great auk had been caught in recent years, despite repeated journeys to Eldey. Hákonarson was keen to find out what had become of the birds—and the two British naturalists had promised to pay him and his crew well. They had to be patient and wait for the opportunity, whenever that might be. Icelanders often, especially in difficult times, refer to life itself as a turbulent sea (*ólgusjór*).

Despite the foul weather, Wolley and Newton spent much of their time outdoors. They did their best to assess the sea conditions, and examined the vessels and landing places of the locals. They rode out along the cliffs, which command a view of the distant skerries and islands (see plate 12). Wolley piled up rocks, making a cairn to serve as a landmark, and, with the aid of his telescope, made a drawing of Eldey and an adjacent rock pillar, getting a good sense of the context in which great auks had once—and might still—lay their eggs. They checked their observations against the "beautiful" map they had purchased in Copenhagen. Newton later remarked: "From our observations, Mr. Wolley and I had reason to doubt whether the bearings of these islands have been correctly laid down."[5]

Our two British intellectuals were understandably "all at sea" concerning the safe operation of open rowboats in rough ocean waves, but they were not much better suited to Icelandic life on dry land. To the people of Kirkjuvogur, they must have seemed nearly as helpless as the great auk on Eldey, with its waddling gait and useless, stubby wings. Slowly, however, things began to make more sense to them, as they took part in meals, shared coffee, and rode on horseback around the countryside with their Icelandic hosts. The extended family at Kirkjuvogur was happy to give their visitors advice and did their best to entertain them. The household included several orphans and indigent elderly people, whom Hákonarson and his wife had taken in and provided for; the couple were well known as "saviors of the infirm and benefactors of the poor."[6] They also had two fine daughters, ages sixteen and six. The younger one, Anna, was later to achieve some notoriety (and

FIGURE 5.2. Eldey (to the right), southwest of Kirkjuvogur. Drawing by John Wolley, 1858. Wolley's cairn is to the left. John Wolley's *Gare-Fowl Books*. (Reproduced by kind permission of the Syndics of Cambridge University Library.)

break her father's heart) when she left the nest and eloped with her lover.

The two British naturalists were also made welcome by the larger Icelandic society. One day a letter arrived from Reykjavík, addressed to Natural Scientist Mr. Alfred Newton: "At a meeting held today by our branch, you have been unanimously elected an honorary member of the Icelandic Literary Society."[7] Founded in 1816, the Literary Society was a prestigious organization. Newton had clearly established his credentials during his sojourn in Reykjavík. No doubt Wolley, too, had made an impression, but in this academic context Newton, the natural scientist from Cambridge, spoke for them both.

Appreciated though the recognition of the Icelandic Literary Society was, Wolley and Newton were compelled to admit that their ambitious, epoch-making expedition was not yielding any results as yet. On May 28, Newton wrote to his brother: "Here we are at one of the nearest places to the Great Auk Islands, and here we have been for a week. . . . What the results of this journey may be, we may not know for another six weeks."[8]

"Found No Birds"

Like much about the great auk, their nesting habits were—and are—unclear. According to Hákonarson, on one occasion a crew rowed out to Eldey at the end of May or beginning of June and saw freshly laid eggs. The birds stayed on the rock, he had informed his guests, for about five weeks. According to an eighteenth-century Icelandic account, the great auks in a colony like that of Eldey "had their nests and eggs in common."[9] The parent birds fed the nestling until it was ready to dive into the sea and fend for itself, but it's not clear how much time that took.[10] Some early reports indicate that the adult birds swam with their young on their backs, a common practice among birds, including loons (*Gavia*).

Eiríkur Magnússon, who had gone to Iceland's East Fjords as Wolley and Newton's emissary to collect specimens of great auks from Hvalbakur island, returned after a month and gave his report. He had ridden clockwise around the island, from Reykjavík, in the southwest, north around Iceland's great central volcanic desert, then south along the east coast to his childhood home in the southeast; the swift glacial rivers along Iceland's south coast can make a more direct route impassable. Once in the east, he hired men to row him out to the skerry. They were unable to land due to rough seas, but Magnússon made them row completely around the island, "quite close enough to satisfy himself that there were no Gare-fowls on it," as Newton later remarked.

Magnússon reported that the local people had no knowledge of great auks ever having nested there, even though the name Geirfuglasker (Great Auk Skerry), not Hvalbakur, appeared on some maps. He was quite thorough in his research. As Magnússon's biographer puts it, he had undertaken "an assignment . . . to find out whether any traces of great auk were to be found on Great Auk Skerry . . . but found no birds."[11]

Though disappointed by Magnússon's report, and disappointed that they could not get out to Eldey to hunt for great auks themselves, Wolley and Newton did not lose focus during their time on the Reykjanes peninsula. The day after they arrived at Kirkjuvogur, Wolley had "picked up from a heap of blown sand, two or three bird's wing-bones (*humeri*)."[12] He was struck with their likeness to great auk bones, and he was not mistaken.

One day they rode ten or twelve miles to the site of the former Church of the Virgin Mary at Vogur, and looked around for more bird bones, without success. Later, back at Kirkjuvogur, they were more fortunate: in the wall of the churchyard there, they found several great auk bones "sticking in the turf, which is used instead of mortar to keep the stones in their places." The bones had been incorporated, inadvertently, into the building material.

But their "most profitable digging" was at the cottage of Sandgerði, some ten miles north of Kirkjuvogur. "Mr. Wolley," Newton later wrote, "one day as he was riding along called out to me that he saw two Gare-fowls' bones lying on the ground. . . . We carefully examined the locality on two other occasions, and found remains which must have belonged to at least eight individuals. Many of them bore the marks of the knife, and nearly all were in good preservation."[13] This was no small matter: No complete skeleton of a great auk had yet been reconstructed, despite the abundance of bones in collections. In the *Gare-Fowl Books*, Wolley fills several pages with details and observations of this skeletal material. The knife marks show these were birds that were eaten.

Some days, the two British naturalists rode south ten miles to the "toe" of Reykjanes, the outermost tip of the peninsula. Facing the grand ocean, they watched the breakers crashing on the cliffs and gazed out toward Eldey with ever-fading hopes of seeing a live great auk. Thirteen years earlier, another British traveler, Charles Cavendish Clifford, had made a foray to Iceland, later reported in comic form in two books written under the pen name *Umbra*. One of his fictionalized traveling companions had promised six different ladies that he would bring them back great auk feathers.[14] He returned home empty-handed, disappointed at failing his admirers.[15]

Wolley and Newton realized they might go home almost empty-handed, too; they would collect no feathers, no eggs or bird-skins, only some discarded bones. It was becoming clear that no expedition to Eldey would be made that summer, as the nesting season was soon—if not already—over. Possibly they could return to Iceland the next summer—it was too early to issue the death certificate of the great auk, they thought—but for the time being, they would have to put aside the big question of the bird's fate.

Bearing Witness

Their quest was not over, however; far from it—in any case, there was no ship back to Britain for weeks. They suspected they were in Iceland at a crucial time, with a unique opportunity to gather first-hand evidence of a species at risk of disappearance. This could be their great contribution to natural science.

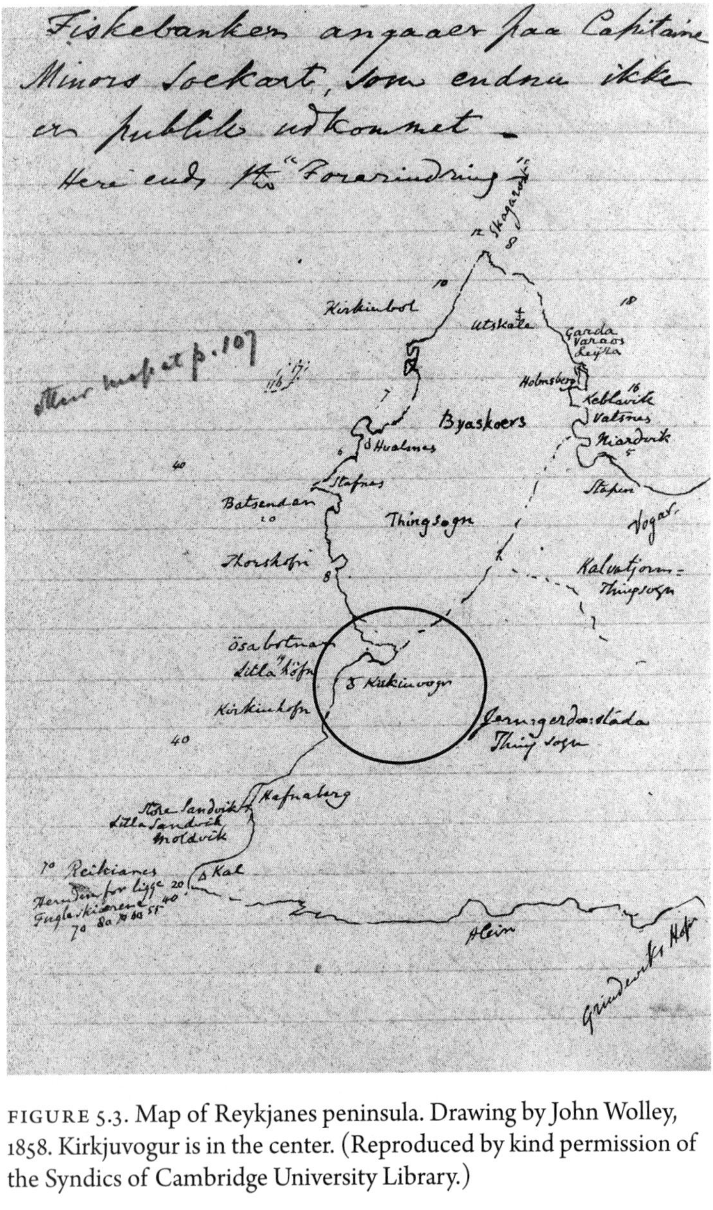

FIGURE 5.3. Map of Reykjanes peninsula. Drawing by John Wolley, 1858. Kirkjuvogur is in the center. (Reproduced by kind permission of the Syndics of Cambridge University Library.)

It was necessary, however, to reboot. Unable to reach Eldey, and having found no sign of the great auk other than its bones, they shifted from a see-for-yourself approach, ideally confirming sightings of living great auks, to striving to shed light on the species' nature by meeting more of the people who had hunted—and eaten—the bird and were familiar with its behavior. The two men arranged to spend most of their time, while stuck ashore, visiting farmsteads and engaging in intensive dialogues with peasants and fishers about their lives and *their* first-hand experience of the great auks of Eldey. They redefined their Victorian pursuit to examine the society around Kirkjuvogur—despite culture clashes, intense socializing, and difficult schedules, they would pay attention to human responsibilities toward other animals, and the existential threat of mortality. From ornithologists, they became anthropologists. This new approach would pay off later, although they did not fully understand it when in the field. Many Victorians were keenly interested in the relations between science, sentiment, and cultural authority, and Newton would later contribute notably on this score.[16] Their interviews would prove to be more valuable than they had imagined.

The Victorian age busily theorized the alignments of reason and emotion, science and sympathy, the subjective and the objective. These alignments were expressed in several forms and on many fronts, ranging from the division of intellectual labor (botany and zoology), the relations of the public and the private spheres, the masculine and the feminine, the allocation of cultural authority, and the challenges of animal protection.[17] Newton was not immune to these currents of thought and practice. His understanding of extinction and animal protection, as Henry M. Cowles has shown, was informed by a particular framing of sentiment and science.[18] In particular, Newton distinguished between two kinds of sentimentalists which, he argued, should not be confused. On one hand, there were "mere sentimentalists" who opposed the killing of any birds at all, and on the other hand, there were those who based their sentiment on "common sense," attending, for instance, to the time and season in which the birds were killed. Newton advocated the latter; birds should be killed only when, and in what numbers, they could be replaced by the population. The protection of animals, he reasoned, should be guided by "a right feeling—a feeling sanctioned by humanity, by Science, and our own material interests," not by the sensibilities of "humanitarians and sentimentalists, whose efforts are sure to be brought to nothing through ignorance and excess of zeal."[19]

Wolley and Newton had many questions about the great auk's status, and, after all, they were in the right place, though their tolerance must have been strained by what they experienced at the various rural homes where they stopped. Iceland's peasant farmers and farmworkers lived in poverty, and that left its mark on their homes, health, and living conditions. In a report some four decades later, a regional physician wrote:

> Physical cleanliness is sadly deficient, and I am of the view that those who rinse off their body a little once a year regard that as an achievement . . . in short, any sense of hygiene is very much dormant, although there are honorable exceptions, and many of them, fortunately.[20]

Hákonarson, their host at Kirkjuvogur, proved to be most helpful. Not only was he the foreman of a "lucky" boat; he was renowned for rescuing mariners from peril at sea. The role of the foreman was to ensure safety, not to compete with other boats or to excel in landing fish. This was radically different from the role of the heroic skipper of a fishing boat in the expanding market economy of the twentieth century, whose main concern was catching the most fish.[21] In his *Gare-Fowl Books*, Wolley captures the essence of foremanship. One local man was said to be "a very able man and a good foreman, so that people liked going with him."[22] The foreman of a fishing boat was the first among equals. But in addition to being a good foreman, Hákonarson was leader of the district council and a settler of disputes; he had a good memory, was popular, and had a good reputation. With the expedition to Eldey more or less officially called off, he agreed to help Wolley and Newton with their "examinations." Some of their "witnesses" were summoned to meet them at Kirkjuvogur, while others they visited in their homes nearby. Sometimes Hákonarson rode with his guests to visit the other farms, and occasionally he contributed to the conversation.

Their first formal interview was with laborer Jón Gunnarsson at Kirkjuvogur. He was seventy years old, born and raised on the south coast of Iceland. He had been to Eldey no fewer than five times. Along with Wolley (and presumably Newton), the interpreter Geir Zoëga was present. Gunnarsson had probably never imagined such a gathering, with himself as the center of attention among learned men. For Wolley and Newton, it was the promising beginning of a long conversation. They continued the interview a few days later.

As a rule, Wolley's *Gare-Fowl Books* make scant reference to a subject's personality, home, or environment. The focus of the interview was all on the great auk. On one occasion, however, the writer notes his surprise at seeing his subject take salt in his coffee "as in Lapland." This may have been a misunderstanding: the well-water in the region was sometimes brackish due to seawater seeping in. Even thirsty horses refused to drink it. The comment is discreetly added in parentheses—as it was not strictly relevant to the matter at hand. But the drinking of coffee is often a focal point of such discussions. On one occasion, it provides an insight into the relationship of hosts and guests: one of the interviewed men, Wolley notes, was "distressed at having nothing but coffee to give us and seemed not pleased when we showed him that we had biscuits in our pockets; 'but it is not for me,' he said."[23] Perhaps he was embarrassed to receive such gifts from British gentlemen.

"Geir" and the Gare-Fowl

Geir Zoëga, in his role as translator, put Wolley and Newton's questions to the subjects and interpreted their answers. Sometimes, on his own initiative, he would ask for more detail, and point out to his employers some aspect of the account given. He became, in a sense, a co-author of the *Gare-Fowl Books*. Perhaps it was his name—Geir, pronounced much like "Gare"—that led to frequent references to the Icelandic name of the bird under discussion: *geirfugl* (*fugl* being a cognate of fowl). The two British naturalists found it amusing that their guide should share the bird's name.

Gare, according to the *Oxford English Dictionary*, means "sharp," or "a spear," or even "a sudden fit of passion," though Wolley had other ideas. Zoëga explained that the word *geir* could mean a stripe, while some of the locals believed that it was a reference to a Viking-age weapon, the *atgeir*, a type of halberd that features prominently in saga literature. Wolley regarded that as an unlikely derivation, because the gare-fowl was a peaceable bird, unlikely to fight, even when attacked. He preferred Hákonarson's suggestion, that the first great auk might have been discovered by a man named Geir and the bird named after him.

In Reykjavík, Konrad Maurer had suggested that in Old Icelandic the word *geir* may have been used of horses, and not men or gods. Later that summer, he wrote to them with an apology.

Geir, he says, is found in the figurative language of poetry, and in personal names like *Þorgeir* (coupled with the name of the Norse god Thor). The learned debate was long and complex. It is not surprising that confusion sometimes arises about the names of birds and men; a recent book on the history of the great auk states, for example, that Wolley and Newton's assistant was "known as 'Geirzoega' (i.e., 'Gareleader')."[24]

Zoëga's biography, published in 1946, gives a detailed account of his work as a tour guide from the mid-nineteenth century: "He was for many years the best-known of all Icelandic guides, and indeed his reputation spread far and wide, both through travel books and oral accounts from visitors to Iceland."[25] The biographer recounts various expeditions made by Zoëga with princes, lords, and other dignitaries around 1856, adding: "In the following two years there were fewer travelers, but in 1859 a huge number of people flooded in, both for scientific research and for pleasure."[26]

One of the people Zoëga assisted was the artist J. Ross Browne (1821–75). Browne wrote this heartfelt tribute at parting:

> A grave, dignified man is Geir Zoëga, large of frame and strong of limb . . . a jewel of a guide, who knows every rock, bog, and mud-puddle. . . . I have traveled many a rough mile with thee, used up thy brandy and smoked thy cigars, covered my chilled body with thy coat, listened to thy words of comfort pronounced in broken English, received thy last kind wishes at parting, and now I say, in heartfelt sincerity, all hail to thee, Geir Zoëga! A better man never lived.[27]

Zoëga's biographer does not mention Wolley and Newton's historic expedition, or the *Gare-Fowl Books* that record their work. But Zoëga spent a considerable amount of time with the

Geir Zöega.

Iceland

FIGURE 5.4. Geir Zoëga. Drawing by Bayard Taylor, 1862. (Library of Congress, Washington, DC. Public domain.)

two naturalists, and kept in touch with them in later years. Perhaps the biographer, in the spirit of his times, saw the bird enthusiasts as less important than the eminent visitors he lists, and dismissed their scientific aims as insignificant. Perhaps there was some embarrassment about the expedition. The Icelanders involved probably felt there was little to say about an expedition in search of birds that were never found, despite strenuous efforts and great expense.

The Way of the Dinosaur

In general, Wolley and Newton's great auk expedition has received remarkably little attention in accounts of foreign visitors to Iceland in the Victorian age. For example, a detailed essay ("With Iceland on the Brain") in an Icelandic journal, focusing on British travelers from 1772 to 1897, makes no reference to them.[28] A 1963 article in an Icelandic weekly magazine, on the other hand, argues that Icelanders "owe a debt of gratitude to the Englishmen Alfred Newton and John Wolley, for their accurate account of the journey of those who saw the last great auk alive and slaughtered them."[29]

In Iceland the phrase, "the last of the great auks," was once used to refer to any lost creature or obsolete object—not unlike the metaphorical use of the word "dinosaur" in English. The phrase had become less common in recent times—it had almost become extinct, like the great auk itself—but now, in the twenty-first century, it is back in fashion both in Iceland and elsewhere, with new significance as one of the parables of the Anthropocene.[30] A recent novel by the French author Sibylle Grimbert, *Le Dernier des siens* (2022), dramatizes the intimate relationship between a nineteenth-century zoologist who witnesses the massacre of a great auk colony and the last great auk, which he manages to rescue, reflecting on what it means to love an endling, something that will never be seen again.[31] Perhaps this captures the predicament of Wolley and Newton as they imagined it.

The great auk's relative the Atlantic puffin (*Fratercula arctica*) is in rapid decline in some regions, including parts of Iceland. As we now know, thanks partly to the work of Wolley and Newton, such a decline indicates that this beloved seabird with the colorful beak may be nearing extinction. At the same time, the bird's image has suddenly become an indispensable souvenir of Arctic tourism, a genuine symbol of consumption—to the extent that stores dedicated to tourism in Reykjavík and throughout the country are derogatorily referred to as "puffin shops." This peculiar phenomenon of "puffin love" has resulted in massive sales in Iceland in recent years of anything puffin themed. Why is that? When a species dies out, it is not only the decrease in biological diversity that troubles us. Stories of the extinction of species, of the deaths of the endlings, the last exemplars, touch people; they are told and retold. Sometimes people are haunted by them, sparking new and exigent questions. In a sense, the periodic, if not constant, presence of the extinct or near-extinct creature in popular culture and the attention it attracts is partly a grieving response of our own species.[32]

Many indigenous communities around the world have preserved institutions and practices that celebrate the mutual relations of humans and birds, as well as other animals now threatened by rapid environmental change. Muscogee Creek Native American communities in the southeastern United States, for example, engage in a series of "sung dances," such as the Feather Dance, iconic of particular birds with special songs: "the men sing one or two verses. . . . They then raise their voices in a high-pitched cry punctuated with birdcalls," resulting in "a beautiful cacophony of birdcalls."[33] These practices have crafted the relations of humans and birds for generations, often in the background of a rebellion against colonialism. But as the Anthropocene advances, both birds and humans face a particular challenge: our collective human-avian existence will need to take on new forms, changing the songs of birds and the dances of humans.

In the Anthropocene, pressing questions arise about where humanity stands in a nature that is subject both to the laws of natural selection and to the whims of human society: Are we inside or outside the aquarium? Young people today—those who will inherit the earth, the biosphere, and the problems caused by previous generations—sense a need for a potent mass movement. Environmental activists seek to convince the public, through a peculiar combination of civil disobedience and science, of the importance of acting immediately to protect nature.

One reason people attune to the threat of extinction is that it exposes the possibility of human extinction, reminding us of our own fate. British writer Jessie Greengrass addresses this idea in her short story "An Account of the Decline of the Great Auk, According to One Man Who Saw It": "If people ask, I tell them it would not be true to say that I feel the loss of the birds apart from the money, except that it is always a little sad when something is gone, because in any loss you can see a shadow of the way that you will be lost yourself."[34] It is pertinent to speak of extinction anxiety.[35]

Wolley's *Gare-Fowl Books* express much the same sentiment. The Iceland expedition of 1858 was clearly disappointing, given its agenda of visiting the breeding grounds at Eldey and finding live great auks. Wolley and Newton had spent enormous energy and considerable money organizing and carrying out a lengthy visit to the places of the last great auk hunters; they returned to

England two months later without even a single great auk egg or a stuffed bird. Was their trip a sheer embarrassment? Not at all, although Newton in his later writings about extinction never flagged that remarkable contract he had drafted and signed concerning Eiríkur Magnússon's search for great auks in East Iceland, in which he and Wolley had expressly requested that Magnússon kill up to eight birds and collect as many eggs as possible.

The results of their expedition could have been much worse if they had insisted on setting off in boats for Eldey, risking their lives and those of their hosts and crews in the heavy seas of the Reykjanes Race. Several of the early polar expeditions were truly horrific in that way, leaving nothing behind but personal loss and bereavement. And yet even here, there were glimpses of hope. British explorer Apsley Cherry-Garrard (1886–1959) chronicled his participation as a zoologist in the fated voyage of Captain Robert Scott to the South Pole. Despite the tragedy of the prolonged expedition, Cherry-Garrard, who was passionate about the emperor penguin, was able to lift his own spirits, at least, years later. He wrote: "And I tell you, if you have the desire for knowledge and the power to give it physical expression, go out and explore. . . . If you march your Winter Journeys you will have your reward, so long as all you want is a penguin's egg."[36]

Wolley and Newton, after all, had their journey—and though they gained not a single great auk egg to show for it, they produced something much more valuable to us, a century and a half on. The *Gare-Fowl Books*, the unique manuscript that records their Icelandic expedition, provides a vivid, almost real-time account of the extinction of the great auk. It is a historic feat. Their 1858 expedition marked the beginning of a long investigation into extinction. Accounts of merchants, hunters, and sailors confirmed that European museums had hastened the critical decline of the great auk population by offering considerable sums of money for bird remains. Immediately after returning to England, Newton concluded: "As to the extinction of the Great Auk, if it is extinct, I think it has been mainly accomplished by human means."[37]

Now, when extinction hangs over the heads of earth's species like the Sword of Damocles, when one species after another is lost forever, the *Gare-Fowl Books* demand to be widely known. So carefully protected that they cannot even be photocopied, these notebooks are scarcely comparable in grandeur to, say, the manuscript known as Cotton Vitellius A XV, which contains the sole copy of the Old English epic poem *Beowulf*, but which has nonetheless been edited and published in many formats multiple times. Or are the *Gare-Fowl Books*, in some ways, even grander than *Beowulf*? As we face a world of extinctions, might not the *Gare-Fowl Books*—the biography of endlings, of the last great auks known to have been caught—prove to be more valuable, ultimately, than this great work of human art?

Interestingly, *Beowulf* is now seen as highly relevant in Anthropocenic terms. Not only does it reflect on the "last survivor" of a tribe, but the latest translation (2021) also has updated some of its terminology. The translator, Maria Dahvana Headley (b. 1977), imagines "the narrator as an old-timer at the end of a bar, periodically pounding his glass and demanding another," pointing out that in the north, close to the places of the poem (and, one may add, to the habitat of the great auk), "icebergs calve into already-brimming seas, and formerly frozen lands reveal the bones and treasures of the dead, melting into mud." What the poem used to refer to, in an earlier translation, as pillage that "emptied the earth of entire peoples" is now rendered as "We existed; now we're extinct."[38] The word "extinction" has not appeared in earlier translations; now it seems right on point.

Clearly, the *Gare-Fowl Books* triggered Newton's radical ideas, launching human-caused, unnatural extinction as a scientific and political object and making it a central aspect of our modern environmental discourse. Might not more familiarity with Wolley's *Gare-Fowl Books* give us a clearer sense of our existential quest in the face of loss—of humanity's effects on the earth and our relations with other species? Perhaps, if only more people could read them, we might learn from the *Gare-Fowl Books* what the last of the great auks paid with their lives to teach us.

CHAPTER 6

"PROUD BIRDS, WITH THEIR HEADS HIGH"

John James Audubon (1785–1851) painted colorful pictures of the great auk that caused quite a sensation in Britain. In 1826, the renowned American ornithologist traveled around that country to raise subscriptions for his masterwork, *Birds of America*, which comprised four hundred and thirty-five pictures of birds, depicted life-size.[1] Later, Newton saw some of Audubon's works, which may have included the great auk. He wrote in a letter in 1845 of one exhibit: "Audubon's 'American Ornithology' is a daub."[2] Elsewhere he notes that "Audubon . . . is a very untrustworthy authority."[3] Contemporary naturalists criticized Audubon's images, which presumably tried to capture "ideal" types, for being falsifications of nature.[4] He never saw a live great auk, let alone a whole colony.

Now, at Kirkjuvogur, in seeking to form an image of the great auk in their minds, Wolley and Newton, like Audubon, were forced to rely on the descriptions and accounts of others, of those who did have first-hand experience of the bird. They asked anyone who had seen great auks to describe them, and they made detailed records of the birds' appearance. Several of

Wolley's examinees, for instance, referred to a white spot between the eyes and the beak, which is a striking feature of most
images of the great auk—the emblem and identifying mark, in
a way, of *Pinguinus impennis*. One man, Wolley wrote, said the
white spot is "as big as a hen's egg; as witness shows on his hand
with the forefinger of the other."

Another of their examinees remarked, however, that people
did not give much thought to the bird's appearance. They hardly
noticed it—it was quite ordinary, just a bird. Was that possible?
One should keep in mind that the great auk hunting expeditions were few and far between by then; hence the knowledge
gained by experiencing the hunt was not passed down reliably
from generation to generation.

Mimicry: "Like This!"

In her 2018 study of mimicry in the *Gare-Fowl Books*, Manchester anthropologist Petra Tjitske Kalshoven develops the concept
of "technique of contiguity," the acting out of another species'
morphology on one's own body.[5] Mimicry, she argues, helps to
amplify the realistic effects of taxidermy, creating the illusion of
life. Wolley and Newton's informants were busy applying the
technique of contiguity, staging a performance to get a sense of
a form of life that was, at least temporarily, unavailable for direct
inspection.

Those who had seen a live great auk said that the bird's posture
was "haughty." While it resembled the other seabirds it nested
with, its behavior was somewhat different. As one man demonstrated, using a stuffed little auk (*Alle alle*), the little auk "keeps
its beak straight and dips it in the water, but the gare-fowl not,
carrying it pointed upward. It is a proud bird." Another man said,
"the bird carries itself like this," then formed a forty-five-degree
angle with his arms, adding that the neck was stretched high.

Several of the witnesses acted out the great auk's movements for Wolley and Newton. Wolley described his host, Vilhjálmur Hákonarson, imitating the bird as it turned its head first to one side, then to the other, and then straight ahead with its neck stretched up, before running across the room with tiny steps. Another witness said he had often seen the great auk: it always thrust its beak forward and swung its head from side to side, "like this!"

In other cases, a dead bird of some other auk species might be used like a puppet to demonstrate. The same witness showed "exactly the action on a dried body (skin and all) of a Little Auk (*Alle alle*) lying in his window." One of the witnesses showed how the hunters approached their prey, tiptoeing across the floor with his arms held wide as if ready to grasp the bird. "As one approached the bird it began to walk a little about its egg, and to turn its head about." With emphasis on his last words, the man held up the carcass of a razorbill, and swung its head from side to side.

When mimicry failed, one of the hunters went so far as to bring a *live* razorbill to show Wolley and Newton, to give them some idea of the behavior of a related auk species. The squirming bird sat on the table, turning its head from side to side and thrusting it forward, as Wolley attempted to make a sketch of it in his notebook. The razorbill tried to fly down off the table, flapping its wings as if it stood up on a cliff. It did not snap at the men, but gave two or three strident cries. This must have been a memorable interview.

Dull-Witted or Dim-Sighted?

A well-known zoology textbook by Icelandic politician and educator Jónas Jónsson (1885–1968) says of the great auk: "The reproductive rate was low, with only a single egg; the bird was dull-witted, flightless, and could only escape by swimming."[6] The

FIGURE 6.1. Sketch of a live razorbill during an interview. John Wolley's *Gare-Fowl Books*, 1858. (Reproduced by kind permission of the Syndics of Cambridge University Library.)

Icelandic term he used for "dull-witted" was *heimskur*, which literally means "one who stays at home"—clearly not true of a bird that traveled the North Atlantic, only staying ashore a few weeks to breed. Why did Jónsson think the bird was stupid?

According to Wolley and Newton's witnesses, the great auk was sensitive to sound and smell; it was necessary to approach it stealthily. One crewmember at Kirkjuvogur, who had extensive experience of great auk hunting, stated that when he went out for the first time, nobody had known how to catch the birds. Another witness said: "They all stand on one rock—and it was easy to get them because they could not run back in consequence of the cliff. The birds ran to the left and the men intercepted them. . . . The bird was quite silent—but then it was seized round the neck."[7]

But the local people thought the bird must be either blind or dim-sighted. As one man reported:

He has said that he had twice seen great auk for himself at sea . . . but only one bird on each occasion. He was surrounded by [little] auks, and describes the bird as it should be—said he had seen a white flap in front of the eyes, and that the bird could not see when it was above water, but appeared to be listening and was startled if it noticed creaking or the splash of oars, and had then dived, surfacing again a long way off.[8]

This informant believed that the "flap" in front of the eyes opened, like a curtain, once the bird was underwater, where it had perfect vision. But another man, having seen a great auk swimming at sea around midsummer, reported that the birds were blind at sea. Others had noticed the bluish-white flap on the bird's eyes; one man who came across a dead bird said that he did not recall it exactly, but that he thought the flap came from under the eyelids. Still others maintained that the bird had no such flap over the eyes, but that on the eyelids was a snow-white spot.

One witness at Kirkjuvogur told Wolley about seeing a great auk from a fishing boat: "When he came about half a mile from Skagen, he saw the bird; . . . the bird followed them for a mile. Witness is sure the bird could not see them. He cannot understand why the bird followed them. It was calm, good weather. The bird did not call."[9] Another fisher also encountered a great auk at sea: "He came a little after them, and they stopped . . . to catch fish. The current brought the boat to the fish. It was good weather and the bird was swimming with the other birds backward and forward."[10]

In cases like this one, the two British naturalists likely knew better than the locals. They were accustomed to handling all kinds of seabirds, alive and dead, and discussing them in detail. They must have been aware that some birds (and also fish and frogs) have a nictitating membrane (from Latin *nictare*, to blink)—an extra, transparent eyelid which can cover the eye to keep it moist and protect it. Biologist Richard Owen, a Cambridge acquaintance, had described the membrane and its function in 1835.[11] But if Wolley and Newton knew the true explanation of the flap in the bird's eyes, they did not mention it in the *Gare-Fowl Books*. Here, their focus was on listening to the local people. At no point do they seem to lecture to their informants.

The Great Auk's Cry

The hunters were generally in agreement that the great auk had an unusual cry, a very low croaking sound unlike the calls of other birds. If humans approached, the bird would cry out, and also if other bird species, such as gannets or razorbills, approached too close in the bird colony. According to one man, the great auk's cry was most similar to that of a puffin—a groan—if it was startled.[12]

Ole Worm could have made notes on the song of the great auk that dwelled for a while in his *Wunderkammer* in Copenhagen in the seventeenth century; he didn't, as far as we know. There are, needless to say, no recordings. Not too long ago, however, German artist Wolfgang Müller had the idea of conjuring up the voices of extinct birds in audio works for radio.[13] The croaking of the great auk had probably originated from deep in its windpipe, Müller thought; he also had to bear in mind that the bird flapped its stubby wings when it called out. Perhaps he reflected on the English name of the great auk. It seems plausible it was onomatopoeic, derived from the bird's low and grating call, a deep squawk: *awk*. Sometime in the Middle Ages, the word "awk" became the root of "awkward," which meant neither upward nor downward but sideways, strange, or against the current; it is now used to mean unwieldy, inconvenient, embarrassing, or problematic. Perhaps medieval people who had some experience of great auks saw an association between this fat, flightless bird and awkwardness.[14]

Müller's *Great Auk* was the first stage of a larger-scale work, *Séance Vocibus Avium* (2008), in which a group of artists re-creates the voices of extinct birds. Müller sought out descriptions of his birds in scientific papers and selected eleven species, including the great auk. Then he summoned a group of artists, and gave them his instructions: "Don't try to be original or funny! Don't try to be yourselves—I don't want to hear you. This is a matter of lending your voice to an extinct bird for a minute or so." The work is accessible online and on CD.[15] Winner of the Karl Sczuka Prize for Works of Radio Art, the work reminds us that organisms that are extinct remain among us—in their own way.

A Peculiar Assembly

While clumsy on land, out at sea the bird was in its element, the hunters Wolley and Newton examined agreed. This was consistent with the information on the great auk provided by Guðni Sigurðsson (1714–80) in a 1770 report copied by Wolley: "Nature has deprived it of that which it has bestowed on other Birds, that is to say Wings; yet it flies with its Wings in the sea as swiftly as flying Birds in the Air."[16] The great auk "flew" swiftly in the sea, making its Latin species name *impennis* (wingless or without flight feathers) a misnomer. Perhaps Sigurðsson was beginning to suspect the principles of the theory of natural selection when he wrote that "nature had deprived" the gare-fowl of what it had bestowed on the birds of the air. The size of wings is relative, in the context of a bird's environment and life-style. Swans and other wide-winged monarchs of the skies could not compete with the great auk in the ocean: They could not dive and hunt with their great wings. The great auk sac-rificed the ability to fly for skill in swimming to chase after, or avoid, fish and seals; its small wings functioned like fins or paddles.

The great auk remained at sea almost year-round, devouring fish of many kinds, and had been spotted swimming in the North Atlantic all the way from Iceland and Greenland to New-foundland. That much Wolley and Newton knew. Little else was certain of the bird's behavior or migration patterns, but Iceland-ic fishers would note their arrivals in the spring. Some said the birds "went in a line as if going to Eldey." The absence of the great auk from the gaze of hunters, naturalists, and ornithologists during most of the year implied that details about its seasonal travels from one colony to another and its social life along the way were largely unknown, at best only a good guess.

Many North Atlantic seabirds are wide-ranging cosmopoli-tans, briefly returning to land to breed between their extensive ocean travels. While some of them are known to consistently appear at predictable times and places, little has otherwise been established about their routes at sea. At their nesting places, seabirds fashion their fuzzy, entangled spaces, often sharing them or negotiating with conspecifics and other species, mu-tually establishing territories at the time of hatching with a polyphonic assembly—presumably much like terrestrial song-birds, according to the analysis of Belgian philosopher of sci-ence Vinciane Despret.[17] The ocean is not a quiet domain, as one might think; it is alive with the sounds of many species, including ocean birds, a roaring fishcotheque.[18]

Recent studies have helped to fill in some of the empty spaces of oceanic life with the aid of modern electronic tracking devices. One of the breakthroughs is a study by a large international team of researchers of seabird migrations in the North Atlantic.[19] The team downloaded massive tracking information from the so-called BirdLife Seabird Tracking Database, focusing on twenty-one species from fifty-six colonies. Analyzing the migration patterns, they made a surprising discovery. Up to five million seabirds regularly assemble at a particular hotspot in the middle of the North Atlantic, the size of France, far away from land. This is an area of lively upwelling within deep ridges in the ocean floor, allowing birds to stop on their way north or south to replenish their energy reserves and, ostensibly, to socialize.

It seems almost certain that the great auk took part in the lively bashes. Some of its surviving relatives and neighbors at key breeding sites in Newfoundland and Iceland, including the puffin and little auk, certainly do. Another recent study (based on GPS-equipped capsules) would support such a conclusion; the great auk seems to have followed a similar trail aided by ocean currents, waves, and wind, gathering strength before continuing to Funk Island in Newfoundland or Iceland's Eldey.[20] The existence of this peculiar seabird assembly at the North Atlantic hotspot has important implications for conservation. Keeping in mind the declining numbers of many seabirds and the threat of extinction, researchers suggest the North Atlantic hotspot should be designated a year-round Protected Area.

Such studies remind us how much can still be ascertained or envisaged about the extinct great auk, using modern technology. For Wolley and Newton in the mid-nineteenth century, interviews with witnesses were the only "tracking devices" available; only by examining the hunters, whether by meeting them at Kirkjuvogur or by calling at nearby cottages, could they imagine the trajectories of the birds.

Skins, Bones, and Feathers

It's important to remember that when Icelandic hunters in the middle of the nineteenth century risked their lives to row out to Eldey and catch great auks, they did not do so out of necessity, to find food, as seafarers had done for centuries. Instead, they were urged to kill the birds by covetous merchants in Reykjavík, who were themselves in contact with obsessed collectors and scientists abroad. The prize was not the meat or oil but the bird's skin, bones, and feathers, which were then stuffed for display or preserved for storage in public museums or private *Wunderkammer*, "cabinets of curiosities." Ostensibly, the last great auks were killed for "science."

The history of the *Wunderkammer* stretches far beyond the Victorian age, originating in the sixteenth century or even earlier. Freely mixing wondrous things, strange animals, and mystic ideas, the cabinet of curiosities appealed to rulers, aristocrats, and members of the merchant class throughout Europe. This was proto-science, without strict disciplinary distinctions and compartmentalizing. Learning and entertainment went hand in hand, playing with the bizarre, the fantastic, and the amazing—all of which were expanded and exaggerated during the Victorian age. Appeals to the exotic, products of colonial expansion and contact, signified the birth of both grand expeditions and comparative anthropology—and massive *Wunderkammer* that grew into modern museums.

Today, cabinets of curiosities are sometimes dismissed as pointless, like stamp collections. In a time of mass extinction, the *Wunderkammer* seems an anachronism. As Cambridge scholar Gillian Beer remarked in 1972: "Whereas in Darwin's time, specimens—whether butterflies or stuffed animals—were mainly seen dead and static in museums, now we respond to movement and flight. Instead of taxidermy we have film. Things seem more alive, more present, when recognized in motion."[21] Most of the heroic collectors of the Victorian age would probably give up in the face of current conditions, with the living world itself seen as a familiar, if shrinking, object. Yet collections and collections of collections are still being created, and some are flourishing. Soon, no doubt, some collections will exhibit genetic resurrections, living (in zoos and aquaria) and dead (in museums), as a way of compensating for the losses and guilt of unnatural mass extinction. The necessary genetic material may be supplied by taxidermied specimens like those so prized by Victorian collectors.

The word taxidermy derives from the Greek *taxis* (arrangement) and *derma* (skin). Taxidermied birds grow dusty and faded. But taxidermy served various functions in the Victorian age, and still does: capturing wild creatures, preserving their remains, and displaying them in our human world reminds us of their (and our) place in nature. In Wolley and Newton's day, bird-skins of various kinds were readily bought and sold, some being more valuable than others. The great auk was, for instance, a good deal more expensive—to put it mildly—than its little cousin the puffin. A good taxidermist would strive to acquire the best bird-skins—undamaged examples, which would make an attractive display—and set out to bring them "to life." In the case of an extinct species, a well-made example provides a captivating glimpse of a world that is gone forever—it almost brings the bird back from the dead. If no exemplar of the right species is available, a facsimile may suffice—an approximation of the bird made using other materials.

The Natural History Museum of Denmark, for example, was recently offered a third stuffed great auk to add to its collection. This bird had belonged to a school in a rural region; it could not remain there due to organizational changes. The Copenhagen museum accepted the donation, but chose not to display it; the bird now lives in a locked cabinet among about a hundred cabinets containing relics of extinct creatures; the wheeled cabinets run on rails along a long corridor. Although to a layperson the difference is not obvious, this great auk is a composite, put together from the parts of several birds—and therefore not authentic. It's a reminder of the blurring of nature and culture, the ambiguous species concept, and the historicity of life-forms. The egg displayed beside the bird on a small artificial rock gives away the fraud. It is covered with large blotches—nothing like the beautiful streaky patterns typical of great auk eggs. No great auk would be deceived, and probably not many ornithologists. Yet the bird and its egg have served their purpose, exciting the curiosity of many schoolchildren and other bird lovers.

Male seabirds tend to have more colorful plumage than the females, but that is not generally true of stuffed great auks. Few who see them can discern any gender difference, and in the last decades of the great auk's existence, there was a frenzied wave of taxidermy of all birds that could be found. But most stuffed great auks are labeled male. Kalshoven suggests that here, as elsewhere, there may have been a gender bias.[22] The idea of "teddy bear patriarchy" has been proposed regarding natural history collections in the first half of the twentieth century—that is, that the selection of objects, aesthetic principles, and methods reflected the gender imbalance in society at the time. From everything we know, then, male and female great auks looked much the same. Yet, one scene on the walls of the Paleolithic Cosquer caves in Southern France is reported to depict two male birds facing off as they contend for the favor of a female; perhaps the imagined scene reflects an endemic gendering in ornithology since at least the days of Darwin.[23]

The rarer the bird, the more people vied to acquire one. Sometimes, however, the bird-skin was so valuable that it became too risky to place it on display. That was the fate of a great auk stored away in the Museum of Natural Sciences in Brussels. Such treasures are like gold bullion locked up in bank vaults—and indeed their value has sometimes been stated in ounces of gold. Kalshoven describes her recent visit to the museum's storage rooms: "the curator immediately located the bird, at the bottom of a tall cabinet with glass doors, together with a few other old mounts and study skins and next to a dodo skeleton—these were the museum's treasures, the curator said."[24]

In his essay on the trade in stuffed great auks, W.R.P. Bourne writes that in the 1970s only two great auks were offered for sale, whose price averaged 377 ounces of gold.[25] That price was almost two thousand times higher than when Wolley and Newton were in Iceland: the equivalent today of about $700,000. One of those two birds had been caught off Iceland in 1830. It was bought at an auction in London in 1971 after a national fundraising campaign and is now in the collection of Iceland's Natural History Museum.

"Stuffing Birds All the Year"

The work of taxidermy—the skinning and stuffing of the birds—was often done by women. Zoëga, the translator, arranged for his cousin, Jóhanna Jensdóttir in Reykjavík, to skin some birds (not great auks, of course) that Wolley and Newton

FIGURE 6.2. Great auk sold at Sotheby's auction in 1971, now in Iceland. (Courtesy of Reykjavík Museum of Photography.)

intended to take home. These skins probably ended up in collections in England or were exchanged for other objects. When Wolley and Newton visited, Jensdóttir told them that she had skinned two great auks while she was living with her mother, before she got married. She skinned them the same way as she skinned any other bird, she said. She recalled that they had tiny wings and a short tail. Jensdóttir's visitors noted that she had an expert touch born of experience. Among the tools and materials she used were knives, hemp, cloth, a clay pot, and a yellow salve to prevent decomposition.

Sigríður Þorláksdóttir, a sixty-six-year-old woman living at Njarðvík, near Keflavík, had often stuffed birds (see plate 13). She showed Wolley, Newton, and Zoëga her method, demonstrating on a common loon (*Gavia immer*), which she opened up under the right wing, as she normally did. The skin was thick and fatty and hard to clean. She stuffed the bird with turf and fine grass—the kind that grew on the sod roofs of Icelandic farm-houses. She inserted pieces of wood to stiffen the head and neck.

Zoëga later returned to interview Þorláksdóttir again, coming back with her account, which he had written down in pencil. In a remarkably precise and vivid narrative, Þorláksdóttir talks about her craft among her other life events:

> She ... had left Keflavík when His Highness the Prince came there ... and she is sure that it was 1834, for that was the summer when she married for the second time, for she has the marriage certificate which is dated July 6, 1834. Her first husband died at Easter 1834, and she thinks she was stuffing birds all the year he died. She recalls that she was out of doors all day in the fine weather, stuffing birds. She says that she would surely not have been away from him so much, had he lain ill in bed—for he lay ill for three years. She does not recall stuffing a gare-fowl before or after that. . . . She thinks that she was stuffing the gare-fowl around midsummer, for the weather was so very good during those days, and the haymaking had not begun.[26]

Zoëga praised Þorláksdóttir for her intelligence and accuracy. He reported that she had been thinking over the questions she had been asked during their earlier visit, and had remembered much more. For her, stuffing birds seemed like a full-time job.

The eggs were often prepared by the same women who skinned the birds. They would make a hole in each end of the egg with a thick darning-needle, then blow out the contents. The eggshells were unusually thick, but still care had to be taken not to damage them. The contents were eaten, either raw or cooked. The shells were sold to merchants in Keflavík, who passed them on to buyers abroad. Sometimes the eggshells were packed in an *askur*, a decoratively carved lidded bowl, and taken on horseback to the nearest town.

There was a lot of rather coarse-textured flesh on the great auk—as much meat as on a lamb. It was regarded as excellent meat; some of the witnesses said it was the best fowl they had ever eaten. As a rule, it was boiled, but sometimes it was roasted. In many cases, a hearty soup was made, in which the bones were also used; otherwise, they were discarded on the refuse heap. Only a very few bones were kept for use in stuffing the bird skins. The guts were also normally discarded. Without doubt, the most profitable part of the great auk was its skin, which fetched ever-higher prices on markets abroad.

The Elusive Brood Patches

While the taxidermists were acutely aware of many of the characteristics of the great auk's body, they failed in one respect, given the account in Wolley's manuscript; they don't seem to have mentioned any brood patches, featherless skin on the underside of birds during the nesting season.

Many birds have acquired brood patches to ensure heat transfer to the eggs when incubating. Some, including close relatives of the great auk, have a single patch, and overwhelmingly the literature has assumed that the same applied for the great auk. Perhaps Martin Martin's 1698 account in *A Late Voyage to St. Kilda* launched the single-patch thesis, referring to a bare spot on the great auk's breast, "from which the feathers have fallen off with the heat in hatching."[27] Now it seems the issue is more complex. Tim Birkhead and his colleagues argue that Martin's claim has been uncritically repeated in later accounts, with one important overlooked exception. In his *Naturgeschichte der Vogel*, German scholar Johann Andreas Naumann (1744–1826) states that male and female great auk "alternate in incubating," "both sexes" having one brood patch "on each side of the belly, as it is the case in *Alca torda*."[28] Observers may have missed the patches, because they would shrink when the skins dried.

Naumann, himself a skilled taxidermist, probably based his account on direct observation—a mounted specimen he prepared from a skin of Icelandic origin is known as "Naumann's Auk." To settle the issue, Birkhead and colleagues carefully studied eight great auk specimens in detail, concluding that "unequivocally . . . the Great Auk had two lateral brood patches." This is not as trivial as it may sound. Assuming two brood patches, the great auk probably incubated in a horizontal posture, but most accounts, based on slight observations during stressful hunting expeditions, imagine a "proud" upright, penguin-like posture: "As far as we know, no one sat and observed undisturbed Great Auks on their breeding grounds and recorded their natural behavior." According to the *Gare-Fowl Books*, one of the crew on the latest successful hunting trip to Eldey, Ketill Ketilsson of Kotvogur, remarked that as they hurried to the colony on the rocks, the two birds they encountered "sat with their heads high, extraordinary to see them."

Birkhead and his colleagues reason that their conclusion has important implications for the understanding of great auk breeding colonies—in particular, incubating horizontally would demand much more space, inviting a reinterpretation of the number of birds nesting on Funk Island, Great Auk Skerry, and Eldey. Wolley and Newton may have failed to raise the issue of the brood patches, taking common assumptions for granted. While often "the eyes of a guest are perceptive," as an Icelandic saying goes, they too have their blind spots. The curiosity of the fieldworker is necessarily relative; the questions asked being dependent on context and social history. "On hindsight," is a common note in later reflections on fieldwork, whether ornithological or anthropological.

CHAPTER 7

QUESTIONS IN, AND OUT OF, TIME

The *Gare-Fowl Books* tell of one bird that was caught alive—and permitted to live. It was on an occasion when twenty-four great auks were caught in one trip to Eldey, probably in 1831. Jón Brandsson (son of the renowned foreman and boat-builder Brandur Guðmundsson) took the bird "for fun."[1] It was around Midsummer Day, said one of the witnesses: "As the grass was growing well around the farm buildings, the live bird was first placed in the grassfield." It sat down, frightened and distracted. The bird had bitten Brandsson on the lip. On the way to shore, the bird showed no fear of the men in the boat.

Brandsson kept it overnight in a locked room, where it "gently knocked at the window." The following day, he tied the bird's beak shut, put it under one arm, and strode off to Keflavik, in case a merchant should be interested in selling the live bird to a buyer abroad. According to another version of the story, also documented by Wolley, Brandsson rode his horse into the village. It was "a great big bird—it tried now and then to get free and snapped at people, but they had put a cord around its neck."

Perhaps rumor had it that some merchant in Denmark was looking for an exotic pet—like Ole Worm with his leashed great auk two centuries earlier. Brandsson commented, however, that the bird "was killed, or at least died"; the fate of the bird's remains is unknown. While the accounts of Brandsson's journey with the bird in the *Gare-Fowl Books* are brief, they present a symbolic, tragicomic image of the tired-out farmhand setting out overland—by horse or on foot—carrying a big seabird under his arm, a bird that silently swung its head from side to side on its way to market.

Sadly, we have few such records of interactions between humans and great auks, with the exception of accounts of slaughtering the birds, which was quick work. Wolley and Newton document only this one case of a captured great auk being kept alive in Iceland, and that was only for a few days. Newton later made a note of another case from the British Isles, citing an article, "The Last of the Great Auks," published in 1848. This bird was found swimming near the shore of Waterford harbor in Ireland. A fisher captured it, "supposing it to be some kind of goose." The man "kept it alive on milk and potatoes . . . then sold it. It was kept for four months. For some time it refused to feed, but at last returned to its diet of milk and potatoes. Later it was fed on fish, but always retained a liking for its vegetable diet, and died, like a good Irish auk, from the result of trying to swallow a too large potato."[2] Around 1750, a great auk from the Farne Islands off the coast of England was kept in the home of a certain John Bacon in Etherstone, Northumberland.[3] The bird was said to be very tame, following its master wherever he went. Clearly, the great auk was usually the "underdog" in the company of humans.

The World of Birds

In many scholarly disciplines, animals other than humans are ciphers—with nothing to say, and no life or relationships of their own. Even taxonomists and evolutionary scientists, since the days of Linnaeus and Darwin, have tended to see animals primarily as exemplars of a species, or types with predetermined qualities and presumed responses.[4] People who have had the opportunity to live with or spend considerable time with birds, dogs, cats, horses, or other creatures, domesticated or wild, often take a different view, thinking instead about the singularity of an individual being and its unfolding life.

An animal's abilities, wishes, or thoughts have generally not been granted much consideration, at least in recent centuries. But there is much to recommend the scrutiny of the lives of animals and their manifold relations to humans. The place of nonhuman animals in social life now demands and enjoys more respect. For many people, the idea of the nonhuman assumes a negative position, establishing an anthropocentric, binary contrast with "us" as the baseline. For some, the terms *more-*than-human or *other*-than-human deliberately negate that negation, restoring a balance that has been missing. All animals, it is widely accepted now, are capable of communication, thinking, pain, and grief—even subjectivity. British anthropologist Tim Ingold (b. 1948) points out that societies are inevitably hybrid, made up of different species, not just conspecifics; human societies, he says, include "all those restless, mutable, roving beings with whose lives our own are necessarily entangled, as much as they are entangled with one another. This entanglement entails a sharing of meaning, of interests and of affects."[5]

Relationships between humans and other animals are of many kinds: They can be based on sympathy and friendship, even respect and equality, but also on oppression, exploitation, and injustice. It is tempting to speak of such relations in classic social science terms referring to relations among humans.[6] The cooperation between eiders and the farmers on whose land they nest in Iceland and elsewhere is close and reciprocal: the wild ducks return year after year to the same nesting grounds, where the farmer provides good nesting sites and protection from predators; in return the farmer is rewarded with valuable eiderdown, taken from the nests in small quantities that do not endanger the viability of the eggs.[7]

Such companionship, if that is the proper term, is not that common, as far as humans are concerned. In their free state, birds are rarely a danger to humans, except when nesting birds perceive a threat, and on occasion they may take defensive action. Birds appear to be attacking humans more than they did in the past, due to our increasing encroachment on their nesting grounds. Birds have been known to fling themselves at the heads of people walking too close to nests, and flocks may swoop threateningly around them in what seems almost like organized "air raids." Cyclists and walkers may be taken aback, even intimidated. One website specifies five thousand attacks by crows.[8]

Bird Biographies

Recently, students of human relations with other animals have begun to zoom in on the biographies of individual animals.[9] This field was pioneered at the beginning of the twentieth century by several ornithologists, notably Oskar and Magdalena Heinroth at the Berlin Zoo. Forgotten for decades, their studies of birds "had the remarkable effect of converting collectors into biologists," paving the road for studies of animal behavior more broadly and launching modern ethology.[10] As Vinciane Despret observes, paying attention to the unfolding lives of birds represented a timely shift from previous practices of studying birds "mainly by killing them or taking their eggs to build collections or to work out categories."[11]

In *The Dance of the Arabian Babbler* (2021), Despret notes that these birds (*Argya squamiceps*) of the Arabian and Sinai peninsulas not only walk more than they fly, but they also engage in complex social activities that tend to puzzle ornithologists, playing, exchanging gifts, and dancing for as long as thirty minutes. The complex sociality of the babblers, she suggests, seems to resist common characterizations of the avian world. Despret argues that to understand the Arabian babbler, it may make sense to turn to the literature of primates for comparison, a radical suggestion in an age obsessed with lineages and species perspectives.[12] In fact, there are striking analogies between humans and some birds. Both speech and birdsong are learned forms of vocalization, acquired during critical developmental periods.[13] Drawing upon her fieldwork in New Guinea, American anthropologist Anna Tsing widens the human-avian gaze, arguing that "allowing bird responses to human projects, as well as the other way around, into social and cultural analysis opens more avenues to consider how science and its alternatives variously shape bird-watching practices."[14]

Estonian biologist Jakob von Uexküll (1864–1944) was one of those who strived to understand animals on their own terms, maintaining that it was vital to fathom how a living being perceives its environment and reacts to it, and that the animal should not be regarded as being like a machine.[15] He was of the view that the animal and its environment were not givens to be taken for granted, but that they were in constant dialogue, like melodies in a polyphonic musical composition, a metaphor he cherished.[16] In the multi-voiced Babel of the biosphere (and in many works of fiction), organisms engage in dialogue across the borders of species. In his groundbreaking book *A Foray into the Worlds of Animals and Humans*, von Uexküll argues that animal bodies are "not defined by their genus or species, by their organs or functions, but by what they can do, by the affects of which they are capable. . . . You have not defined an animal until you have listed its affects. In this sense there is a greater difference between a race horse and a work horse than between a work horse and an ox."[17]

The detailed study of the lives of sparrows by Margaret Morse Nice (1883–1974) challenged the prevailing tradition in the 1940s, suggesting that often what is considered as territorial behavior includes display and the expression of presence, sometimes quickly changing from one mode to another, from the company of a few birds to that of extensive swarms.[18] Interestingly, the recent focus on the biography of the individual animal echoes that of Philip Henry Gosse, the Victorian master of aquaria. Gosse had noticed "distinctness" in both birds and fish:

> Doubtless this individuality would be much more generally perceived, if our observations on animals were not so loose and cursory as they usually are. . . . But bearing in mind that records thus obtained of the manners of animals are properly biographical,—belonging to the individual more strictly than to the species,—it manifests that these must be the foundation of all our generalization.[19]

Arguably, however, as Despret observes, focusing on distinctness in the fashion of Gosse, in the confined spaces of both aquaria and aviaries, is liable to yield skewed observations that have little to do with ethological realities outside the tanks and the cages, exaggerating territoriality and what supposedly comes with it, including competition and aggression, beyond proportion.[20] The experimental conditions not only demonstrate the behavior observed, they also *produce* it.

Wolley and Newton were obsessed with the great auk. Was there any way to understand this bird that so captivated them, any more than any other wild creature? But while they meticulously tracked the biographies of eggs and stuffed birds, in Victorian style, tracing their trajectories from places of capture to museums and private collections, they were not at all interested in the life histories of individual birds. Their focus was on lives out of time, with no possibility of observing birds prior to their death; biographies of living birds were not yet on the agenda. Wolley and Newton were no less intelligent observers than the Heinroths, Nice, von Uexküll, and Despret, but the urgent avian quest has been radically redefined since their time. They would be surprised to hear us speak of the *Gare-Fowl Books* as the biography of the last great auks, the endlings of the species. Such was not their intent, and the biography they produced is spotty and inconclusive; but it's the best we have for this bird that went extinct before it could be better known.

Parting Days

In order to tell tales of birds and humans—not least if they are dramatic or establish a landmark, even the end of something—chronology is a crucial factor. When does the great auk find it convenient to nest? When do fishing-boat owners and merchants go out to hunt them? When did the bird stop nesting on the Great Auk Skerry, where the massacre occurred in 1813? When was the last time a great auk was seen on Eldey?

In her works on Icelandic history from the settlement in the ninth century to the nineteenth century, Danish anthropologist

Kirsten Hastrup (b. 1948) explored the sense of time in the evolving social world of the Icelanders and the natural world they inhabited:

> The settlers arrived on an empty island, and their creation was of immense proportions, also as far as the environment was concerned. Time and space were socialized, named, and historicized, in ways which marked the physical space by culture for generations to come.[21]

Some of the chronological markers that the Icelanders used were familiar to Wolley and Newton: the Feast of St. John the Baptist (Midsummer Day) and Candlemas (February 2), for instance. Others were less easy for the two British naturalists to comprehend, such as the *fardagar*, the Flitting Days or Parting Days, an important moment in Iceland's annual calendar and a precursor to the Closing Day of the fishing season. The Parting Days—the first four days of the seventh week of summer (between May 31 and June 6)—were the only time of year when tenant farmers could move from one farm to another, like flitting birds. As it happens, the Parting Days were also a major referent for the flightless great auk. While they were not quite as punctual as the Icelanders, the birds would normally arrive on the skerries to lay their eggs more or less on the official Parting Days.

Other matters of chronology, in this remote region, were also not easy for the visitors to grasp. Some of their witnesses did not know how old they were: One farmer was "aged not quite 40 but does not remember, has not looked in church book." Others could not say when they had killed great auks, or when they had sold them—and least of all could they provide exact dates for significant events in the history of the great auk. Often, they referred to other historic occasions. One states that eighteen or nineteen years earlier, he had come across a dead great auk on the beach: "He remembers that they were baiting lines for skate, which takes place in June." Another recalled going on a memorable hunting trip "when twenty-four great auks were caught," and a shop worker remembered the hunt had been "the same year when the cholera was worst in Hamburg."[22] One man referred to a measles epidemic of 1846, and some mentioned a visit made to Keflavík in 1834 by Crown Prince Frederick of Denmark and Iceland.

But one date was quite easy to establish precisely. When did the bird stop nesting on the eponymous Great Auk Skerry near Eldey? One man remembered exactly: the volcanic eruption that caused the skerry to disappear beneath the waves started on the day his brother died—7 March 1830.[23] A recent work on medieval Iceland juxtaposes its bloody feuds, detailed in the sagas, and the "extra-social menaces" of the age, pointing out that the saga writers were strangely silent on the latter, especially eruptions that they must have witnessed.[24] The disappearance, in contrast, of the key site of great auk hunting in the nineteenth century was hard to miss and forget.

"Entirely Covered with Birds"

In England, when planning their expedition to Iceland, Wolley and Newton had seen frequent references to this Geirfuglasker, or Great Auk Skerry, in the southwest of the country. Not to be confused with the one in the East Fjords, better known as Hvalbakur, to which they had sent Eiríkur Magnússon, this Great Auk Skerry was quite near Eldey island; the whole archipelago, in fact, was labeled Fuglasker, or Bird Skerries, on their map. As soon as they arrived in Iceland, however, they learned that it was too late to visit this historic nesting ground of great auks; most of the skerry had sunk into the sea in a volcanic eruption in 1830. Instead, they asked acquaintances who had access to old books and papers to scan them for useful information about this vanished great auk habitat.

Konrad Maurer, who had traveled with Wolley and Newton aboard the steamer from Scotland, uncovered what may be the most eye-opening document. From Reykavík, he wrote a letter:

> To John Wolley Esq. Reykjavik, entrusted to the care of Dr. Jón Hjaltalín, Landsphysikus. Being on the eve of my departure for a tour in the interior, I would like to leave a letter for your hands, spelled in the worst of your queen's English, to the care of our mutual friend Dr. Jón Hjaltalín. I send you thereby a copy of the passage in Wilchins Máldagabók [Book of Cartularies] referring to the Geirfuglasker.[25]

This Book of Cartularies from the mid-eighteenth century, recorded by a farmer from Kirkjuvogur, describes the skerry as being "the size of a grassfield to fodder one cow"—or roughly one acre. It was, he says, "entirely covered with birds."[26] One of the witnesses Wolley and Newton interviewed said that the skerry had been just like Zoëga's brown wide-brimmed hat, "high in the middle with low land all round."[27] The dense nesting colony may have numbered hundreds of great auk. Obviously, this would have been a valuable economic asset.

Indeed, the Icelandic *Register of Estates*, compiled in 1703, specified that this Great Auk Skerry belonged to three nearby churches. Since at least the fourteenth century, half of the skerry was the property of the Church of the Virgin Mary in Vogur. Newton comments, however, that "most likely" some hunting grounds were "left to reward the bold adventurers who resorted thither," some foremen or households staking their claims on promising sites: "Two . . . birds were obtained by Stephan Sveinsson of Kalmanstjörn, whom the good people of Kirkjuvogur seem to look upon as a kind of poacher on what they consider their rightful domain."[28]

In the earliest Icelandic written texts, the famous Sagas and Eddas from the thirteenth century, no reference is made to the great auk. Great Auk Skerry is first documented in the cartulary of Kirkjuvogur Church in 1397.[29] The earliest reference to a hunting expedition to the skerry is in the Skarð Annals, recording that in the summer of 1639 four vessels went there, of which two foundered, sinking with all hands. Bird-hunting at Great Auk Skerry was indeed a dangerous enterprise, as witnessed by *Travels in Iceland*, the report of a Danish-funded expedition in 1752–57:

> The sea is . . . invariably turbulent and is sucked to and from these skerries, forming a vortex around them, especially in calm weather. . . . Men of south Iceland, who venture to go there in calm seas, sometimes hunt Gare-fowl. It is not possible to land on the skerry, however: One of the crew must jump ashore with a rope; and when they depart again he must often be hauled up into the boat out of the sea.[30]

Even though forays to Great Auk Skerry were a thing of the past, several local men had clear memories of hunting there, from which lessons might be learned. An account written down in the mid-1800s stated that "only in one place on the skerry was it possible to land, by a certain cliff." One man "went up with a rope, and that man had to know how to swim. Then the boat lay at anchor as close as possible, and all that had to be taken ashore, and back again, was transferred on ropes."[31]

The method is clearly depicted in a dramatic picture in a late-eighteenth-century manuscript, which shows two boats at anchor and three men on the skerry, along with dozens of great auk. Konrad Maurer had advised Wolley and Newton to look for the picture in Reykjavík, and they were not disappointed upon seeing the "curious document." Newton commented later: "Two boats are seen, anchored with larger stones. . . . In one of these are seated three, and in the other two men, waiting the return of three comrades, who are on the rock, hunting what appear to be Gare-fowls, of which upward of sixty are represented."[32] The picture, which has often been reproduced, shows a small shed (actually two sheds combined), probably built to shelter crew that might be left behind because of heavy seas or a sudden change in weather.

Farmers, then, could rely on finding great auks on Great Auk Skerry; sometimes they filled their boat with birds and eggs. The hunting was said to be best when the spring was rainy. In the late 1820s, twenty-seven great auks were killed there on a single trip.

The men described to Wolley and Newton the glow of the volcano and the billowing smoke during the night of the eruption at Great Auk Skerry in March 1830. Within a week, most of the skerry had sunk, although the eruption lasted for two years. Catches on nearby fishing grounds were thinner than usual. The fire could be seen throughout the night, and sometimes there was "great smoke." Vilhjálmur Hákonarson was "sure there was fire in the sea for two successive seasons."[33] Sometimes the smoke could be seen from Reykjavík; some people said they detected fire from that far.

In the twentieth century, Icelandic geologist Sigurður Þórarinsson (1912–83) managed to establish sources testifying to a number of eruptions in the area since the early 1100s. Quite possibly, several of these early eruptions affected great auk habitat; the birds probably adapted by moving, for breeding, from one skerry to another nearby. Consulting Newton's summary of his and Wolley's findings, Þórarinsson argued that the 1830

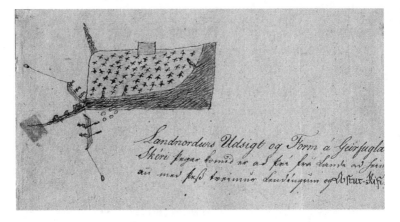

FIGURE 7.1. Hunting of great auk at Great Auk Skerry. (From Landsnefndin 1770–71, manuscript Lbs 44, fol. 71v. Courtesy of the National and University Library of Iceland.)

eruption at Great Auk Skerry "proved decisive for the flightless bird, the great auk, as . . . the eruption and the earthquakes that followed it destroyed its only remaining nesting place at the time."[34]

"Never, Never, Never"

According to the Book of Cartularies, "some withered human bones" were once found on this Great Auk Skerry. On another occasion, "three men had been known to have supported themselves on the rock by eating sun-dried birds and drinking rotten eggs for half a month before they were taken off."[35] This sounded convincing to Wolley and Newton. Perhaps the sheds on the map of Great Auk Skerry were built after these terrifying events.

Other testimony presented to the two British naturalists must have been more trying, as it verged on the fantastical. One informant in his eighties told Wolley that one summer in the fifteenth century, the son of a farmer at Sandgerði had been left behind on Great Auk Skerry. When men rowed out to the skerry the following summer, he was found there alive, though barely surviving.[36] Another man corroborated the main points of the story, and Zoëga said that he was also familiar with the tale. Jón Árnason, whom Wolley and Newton met in Reykjavík, recorded several versions of this dramatic tale in his collection of Icelandic folk and fairy tales, compiled between 1854 and 1861. The version "told by an old man . . . who had long been a fisherman at Keflavík" ascribes the auk hunter's survival to help from an elf woman—who exacted a horrible payment when he later repudiated her.[37]

The old story had an impact: belief in elves and other supernatural beings was common in Iceland (as it still is).[38] Many farmers and their workers became reluctant to venture out to Great Auk Skerry, although for centuries it had been one of the principal breeding colonies of great auk off Iceland. An old man examined by Wolley and Newton told them emphatically that it had "never, never, never" crossed his mind to row out there to hunt. Anyone who stepped ashore on the skerry could not help recalling the story of the revenge wrought by the elves who lived there. Could the daunting impact of the folktale have been a contributing factor in the great auk's survival into the mid-nineteenth century?

The spring that the eruption began at Great Auk Skerry, the *Gare-Fowl Books* recount, a bird-hunter and his crew found dozens of breeding pairs of great auk on nearby Eldey island. Two years later, they made two expeditions to Eldey, catching eight birds on each occasion. The following spring, they caught twenty-four. One witness said that the birds bit his arms and legs and swung their heads around without letting go. One bit him on the cheek. He himself killed twelve birds: first he held seven birds in his arms, then laid them down and caught another five. He tied the half-dead birds together by their feet, looped a rope around them, and then they were hauled aboard the boat.

These were new conditions for the bird-hunters, and far more hazardous than those they had known before. The farmers had never, before the eruption, rowed out to Eldey in search of great auk. It was regarded as impossible to land a boat there; then they heard that Dutch sailors had been seen at the island. They had sailed in close, then gone ashore in a lighter. This led the Icelanders to follow their example. A local hunter wrote the following description of Eldey:

> Eldey is a large rock in the sea west of Reykjanes. . . . It is mostly vertical cliffs with few ledges; on the east side it is possible to land by a projecting rock in calm seas, eighty square fathoms. . . . There is a lot of slime on the rock so it is dangerous to climb up onto it. It is about two fathoms high at low tide, which is the best time to land there. There is a great flux there and a maelstrom. Many birds are found on the rock mentioned here and on a large ledge higher up which is inaccessible.[39]

"It Has Surely Moved"

The year 1830 was a significant marker in the calendar that Wolley and Newton established; it marked the great divide before and after the eruption at sea, when both birds and men were thrown into confusion. It was in 1830 that Eldey took over as the prime breeding ground for great auks, possibly the last one, the site of a historic extinction—a reputation it has kept to this day.

After Great Auk Skerry had largely sunk beneath the waves, much of the great auk population had found itself homeless, as Newton put it later.[40] While some of the birds must have taken refuge on Eldey, others were likely to have gone to Greenland, said one of the men Wolley and Newton interviewed. The local people attributed the change in the birds' habits to a failure of the capelin, a small fish on which the great auk fed. They speculated that great auk might still be caught farther north along Iceland's west coast, or in the east of the country.

According to a report sent to Wolley and Newton, "this bird is not easily identified, and though it is not longer seen in the Eldey islands it has surely moved to another outlying skerry, where it has remained without being scared off. . . . Should I later receive evidence from more men who claim to have seen it in recent years," writes the author, "I shall make a report upon it."[41]

Was there any reason to fear its disappearance from Eldey? Was this not the natural way of a bird's life? Wolley and Newton's witnesses do not appear to have had any real concerns—although some of them had come to suspect that the bird's existence off southwest Iceland might soon come to an end. The people who described great auks to the two British naturalists were not talking about an extinct bird. On the contrary. Their statements were all in the present tense: "The bird flaps its wings as it moves about," "it runs away from one to the left, like auks do." There is no past tense here; no sadness, no nostalgia.

One witness pointed out that during Brandur Guðmundsson's second trip to Eldey, in 1835, he came back with a catch of eight great auks, and only two birds had escaped the hunters; yet twenty-four birds were caught there the following year. Thus, there appeared to be no reason to fear for the birds' overall survival.

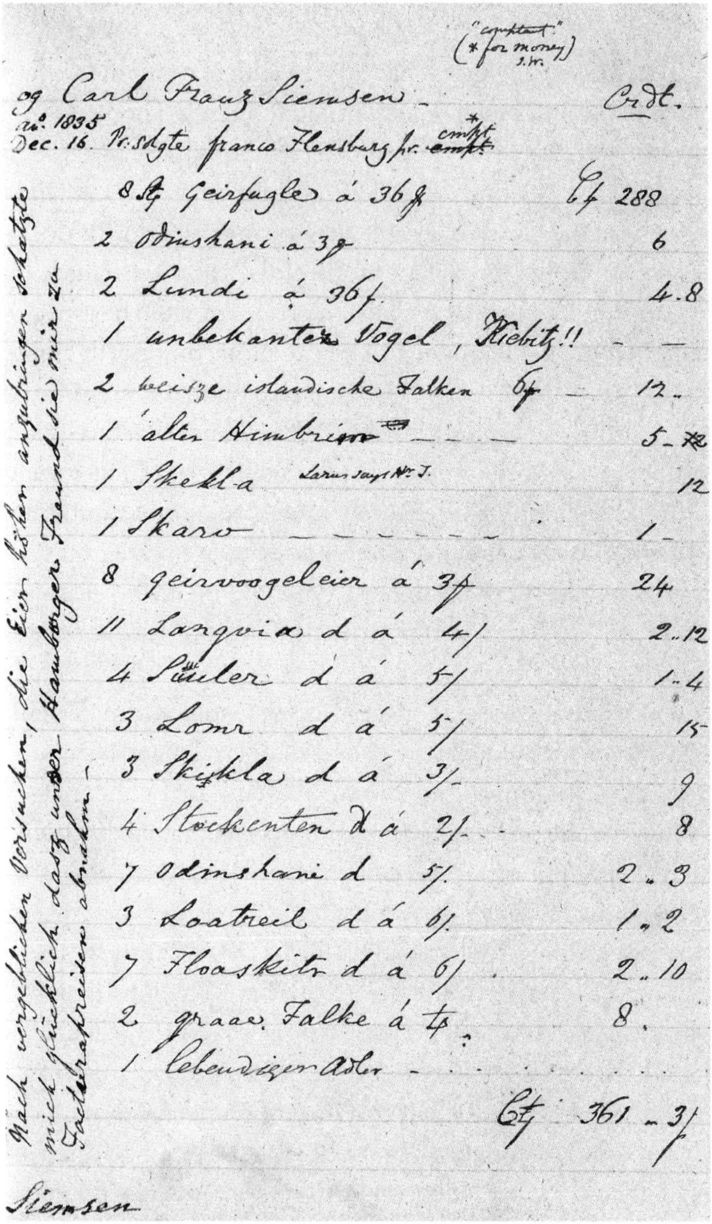

("contant")
(* for money)
s. kr.

og Carl Franz Siemsen _____   Crdt.

nº 1835

Dec. 16. Vrschgte franco Flensburg fr. *empf*

8 Stg Geirfugle à 36 ß   6 f 288

2 Odinshani à 3 ß   6

2 Lundi à 36 ß   4.8

1 unbekanter Vogel, Kiebitz !!   — —

2 weisze islandische Falken 6 f —   12 —

1 alter Himbrimm ___ _ _ _   5 — *

1 Skekla   Larus says Nr. J.   12

1 Skarv _ _ _ _ _ _   1 —

8 geirvoogeleier à 3 ß   24

11 Langvia d à 4 /   2..12

4 Säuler d à 5 /   1–4

3 Lomr   d à 5 /   15

3 Skifkla d à 3 /.   9

4 Stockenten d à 2 /   8

7 Odinshani d   5 /.   2..3

3 Loatreil d à 6 /   1..2

7 Floaskitr d à 6 /   2..10

2 graae Falke à 4 ß   8 .

1 lebendiger Adler _   ·.

                                            Cf 361 ..3 /

Siemsen

FIGURE 7.2. Store document of merchant Carl F. Siemsen, 1835.
Copied by John Wolley and Alfred Newton, listing eighteen species of
birds. John Wolley's *Gare-Fowl Books*, 1858. (Reproduced by kind
permission of the Syndics of Cambridge University.)

Wolley gained access to the accounts of the merchant who sold Guðmundsson's eight great auks to collectors. According to these records, the merchant sold many birds that year, including eleven common guillemots (*Uria aalge*), seven red-necked phalaropes (*Phalaropus lobatus*), and four falcons (*Falco rusticolus*). The eight great auks were the most valuable. Eight great auk eggs were also purchased, one of which was the egg later acquired by John Wolley.

Sometimes the hunters took nets with them to catch quantities of smaller species along with the prized great auk. The birding nets were generally about twenty-four feet in length: one end of the net was held fast, while two men held the other end as they walked about the rock scooping birds into the net, up to two hundred at a time. These were genuine massacres. Birds were a resource—whether as food or to sell to collectors—regardless of how rare they might be. Wolley does not pay specific attention to this issue in the *Gare-Fowl Books*, although on his return to England, Newton would address the subject. Theories of population collapse and bird conservation were in their infancy in 1858.

CHAPTER 8

THE LATEST SUCCESSFUL TRIP

Wolley and Newton often asked their informants about that "trip when foreman Hákonarson got two great auks"—as the locals generally put it. People remembered it reasonably clearly, and most of the crewmen were still alive. Listening to their accounts, Wolley and Newton came to realize that several years had passed since any of the birds had been "captured." Their witnesses would refer to the latest successful trip—not necessarily to the "last of the great auks"—as if the infamous 1844 hunt had not been unusual in any way, and more birds might well turn up next time. Icelandic, it is important to note, makes no clear distinction between "last" and "latest," both would translate as *síðasti* or *seinasti*. The two terms are normally considered identical, although *seinasti* seems to be slightly more formal and more common, possibly closer to "the last."[1] Wolley does not specify which Icelandic term his informants used.

The reader of the *Gare-Fowl Books* perceives a rising suspense in Wolley and Newton's examinations—it's almost a thriller. *When were the last of the great auks killed, and where? Who killed them? And what happened to the birds?* As time went on, Wolley and Newton came to suspect that the great auk had not just flitted away to a new habitat in Greenland or elsewhere in Iceland, like tenant farmers shifting their lodgings on the Parting Days. Yet there is no trace of judgment or accusation in their writings. They had, admittedly, hardly a leg to stand on: Had they not personally commissioned Eiríkur Magnússon to go to the East Fjords to catch great auk for them—eight birds or more, if possible, and as many eggs as he could?

On some occasions, two or more informants were "examined" together, and all were asked to recall the latest successful trip in as much detail as possible. Their narratives are recorded in chronological order, scattered throughout the *Gare-Fowl Books*, but mostly concentrated in the first two volumes. The witnesses had difficulty remembering exactly when the expedition had taken place; most guessed it was ten or twenty years ago. With some hesitation, Wolley and Newton concluded it was in 1844. That date was partly based on the argument that a farmhand who had taken part in the hunt had been employed at the time by Guðni Hákonarson (brother of Vilhjálmur the foreman); as the man had left Guðni's employ in 1844, the trip could not have taken place any later than that. A merchant also stated that he was in Keflavík when the great auks were sold: it was the first year that speculator Christian Hanson was in the village, and he purchased the two birds that were caught—and that had been in 1844.

It was an eventful year. The Annals of Suðurnes report (in typically staccato style): "Ship lost . . . in March. . . . Storm and heavy seas April 2, so boats were washed up and broken. Ship lost in the spring. . . . Three men drowned. . . . Fifty sheep drowned by high tide."[2]

Latest Crew That Got Great Auk

After a number of "examinations," Wolley and Newton decided they were able to determine who had been in the crew when the latest great auks were caught. They concluded that fourteen men had gone on the trip and listed them, together with their ages in 1844, their occupation, and the name of the farm on which they resided, if known. "Last crew which got *geirfugl*," writes Wolley in the *Gare-Fowl Books*, "must have been as follows" (the list has been slightly amended in accordance with available genealogical information):[3]

> Vilhjálmur Hákonarson, farmer, foreman, 32
> Eiríkur Ólafsson, Vilhjálmur's farmhand, 32
> Guðni Hákonarson, Vilhjálmur's brother, farmer,
>   Kirkjuvogur, 28
> Gunnar Halldórsson, farmhand, Kotvogur, 19
> Jón Bjarnason, farmhand, Hvalsnes, 41
> Jón Brandsson, farmhand, Kirkjuvogur, 41
> Jón Eyjólfsson, Vilhjálmur's farmhand, 32
> Jón Gunnarsson, farmhand, Kirkjuvogur, 56
> Jón Þorsteinsson, farmhand, Garður, 33
> Ketill Ketilsson, farmhand, Kirkjuvogur (later farmer
>   at Kotvogur), 21
> Ófeigur Hendriksson, farmer, Kirkjuvogur, 38
> Ólafur Jónsson, deceased, Vilhjálmur's farmhand,
>   a young man
> Sigurður Ísleifsson, farmhand, Hafnir, 25
> Sveinn Þorvarðarson, deceased, farmhand, 35

Wolley did his best to work out what role each member of the crew had played, and he drew up a family tree to clarify their relationships. Ketill Ketilsson and Vilhjálmur Hákonarson had

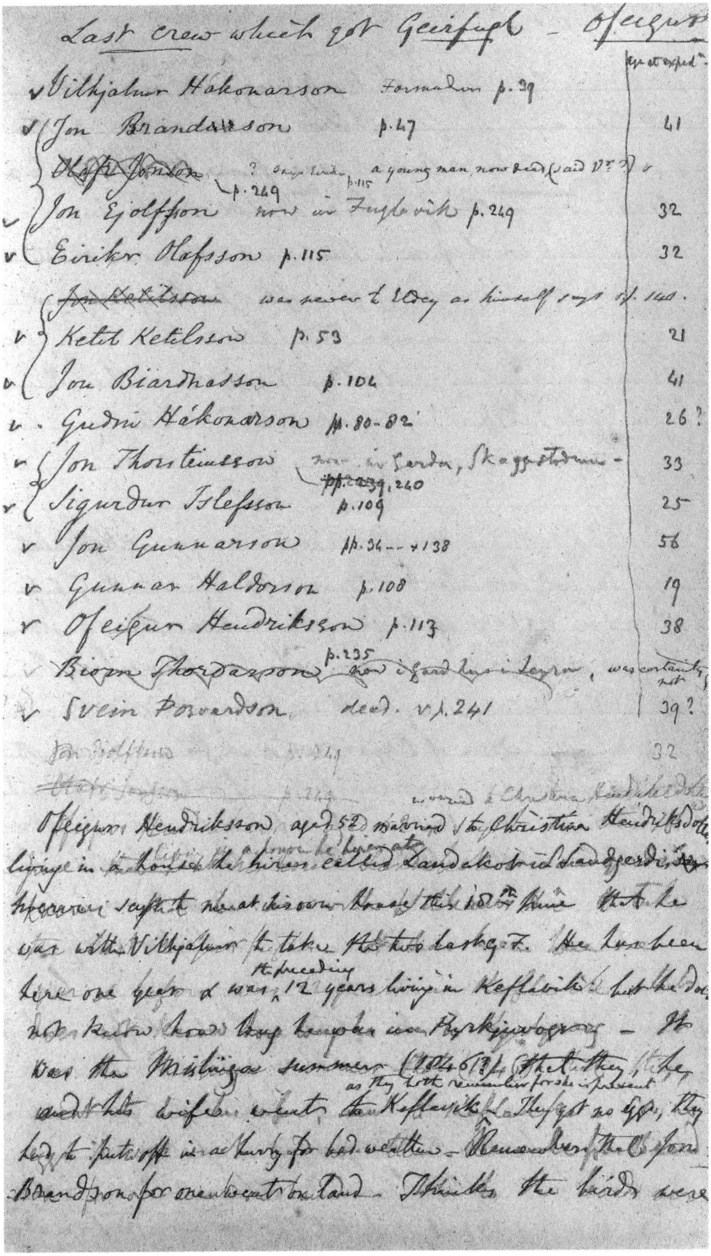
Last crew which got Geirfugl – Ofeigur
Vilhjalmr Hákonarson Formadur p. 39
Jon Brandsson p. 47 41
Jon Gjolfsson now in Fuglevik p. 249 32
Eirikr Olafsson p. 115 32
Ketil Ketilsson p. 53 21
Jon Biarnasson p. 106 41
Gudni Hákonarson pp. 80–82 26?
Jon Thorsteinsson 33
Sigurdur Islefsson p. 109 25
Jon Gunnarson pp. 34 – + 138 56
Gunnar Haldorson p. 108 19
Ofeigur Hendriksson p. 113 38
Sveir Þorvardsson dead. v p. 241 39?

FIGURE 8.1. The last crew that got great auk. John Wolley's *Gare-Fowl Books,* 1858. (Reproduced by kind permission of the Syndics of Cambridge University Library.)

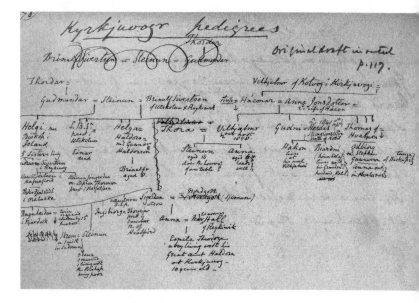

FIGURE 8.2. Kinship relations at Kirkjuvogur hamlet. John Wolley's *Gare-Fowl Books*, 1858. (Reproduced by kind permission of the Syndics of Cambridge University Library.)

been brought up together (Wolley incorrectly records them as being half-brothers). Young farmhand Gunnar Halldórsson (aged nineteen) and Hákonarson were half-brothers. Two of the oarsmen were local farmers, while the rest were farmhands—most of them employed by Hákonarson and his brother Guðni at Kirkjuvogur. Jón Gunnarsson was by far the oldest in the crew, aged fifty-six, and had made more expeditions to Eldey than the others. By the time of Wolley and Newton's visit, only two of the crew were deceased. One of the men they examined was one-eyed. Another had had leprosy at some point.

Wolley may have been confused at first when drawing genealogies, not realizing that Icelanders invariably used first names and that family names were rare. Most people were identified by patronyms (as today); thus Vilhjálmur Hákonarson was the son of a man named Hákon, and taxidermist Sigríður Þorláksdóttir was the daughter of a man named Þorlákur.

Seamen's Tales

The latest successful trip to Eldey was made on the initiative of Reykjavík merchant Carl F. Siemsen, although no formal agreement had been drawn up in advance. In the mid-nineteenth century, collectors around the world were keen to acquire great auks, and through merchants and other intermediaries they sent out hunters to find them. If such a trip were successful, each crew member could expect to be well rewarded. For one trip, each might gain as much as a farmhand could earn for the entire haymaking season.

Hákonarson observed the weather prospects and sea conditions for several days before deciding to set off. Much was at stake, but he would not take unreasonable risks. An experienced mariner was said to have stated that "after northerly weather had gone on for a long time, and a steady land breeze was blowing, it was all right to set off, and it was taken as a sign when the sea was calm at Stafnes [near Sandgerði] that it was safe to land." The time had come. The breeze was steady, the sea was smooth, and the Parting Days had arrived: the great auks had probably already laid their eggs on Eldey island.

Around six o'clock in the evening, one day between May 30 and June 5 (in 1844 those were the Parting Days), the crew pushed Hákonarson's eight-oared boat afloat. They had to make allowance for the tides, taking account of the landing conditions expected at Eldey twelve hours later. Everyone aboard the boat was aware that it was sometimes necessary to abandon the attempt to land, due to high waves around the island. The boat was about thirty feet long—it was regarded as inadvisable to try to cross the Reykjanes Race in a larger vessel—with fourteen men aboard.

Kirkjuvogur itself was sometimes seen as a landing place of last resort: it is the only more-or-less reliable landing place on the whole coast of the Reykjanes peninsula for the sixty miles or so that stretch between the towns of nearby Sandgerði, to the north, and Þorlákshöfn, which lies around Reykjanes on the south coast. During the age of the rowing-boat fishery, when vessels were frequently lost at sea, the landing place at Kirkjuvogur often played a life-or-death role. But with the right skills, taking a boat in and out of Kirkjuvogur was not risky.

Interestingly, the crew do not seem to have mentioned their exact place of departure. It may not, in fact, have been Kirkjuvogur. One of the women interviewed, Guðrún Jónsdóttir, a widow aged fifty-nine, states that she "thinks" the latest expedition left from Kalmanstjörn, a farm farther south.[4] If she is correct, the crew from Kirkjuvogur would have taken their oars and food on horseback to the landing place, about twelve miles. Perhaps women were more attentive to such details, given the frequency of losing a husband or son at sea.

Wolley and Newton do not mention any women who went fishing. This is not surprising, given the context of great auk hunting and the gendered world of Victorian England. The great auk crews known to them were all men—and, apparently, the taxidermists were all women. It should be kept in mind, however, that women's participation in fishing was far more common in Iceland than the *Gare-Fowl Books* indicate.[5] Sometimes women took the role of foremen in the rowing boats. One of the renowned foremen on the southwest coast at the time of Wolley and Newton's expedition was a woman, Þuríður Einarsdóttir (1777–1863), usually called "Þuríður the foreman." It was said she never lost a man to the sea.

FIGURE 8.3. Landing place for boats on Reykjanes. (Photo: Magnús Ólafsson. Courtesy of National Museum of Iceland.)

The crew of the latest successful trip had loaded rocks into the boat for ballast, before pushing the vessel out. They were dressed in unwieldy oilskins of ewe-leather, with leather footwear to keep their feet dry when seawater washed into the boat. They sat apprehensively, each beside his own oarlock, and recited a traditional prayer. Except for the foreman, Hákonarson, who steered, they took turns at rowing. Those who were resting would drink some water or eat the food packed for them by the mistress of Kirkjuvogur.

A change in the weather was not all that they feared. The same year the last great auks were caught on Eldey, howling supernatural beings had made an appearance not far away.

In his private journal, which Wolley and Newton did not have access to, a local man wrote of a fishing trip made by ten men in an eight-oared boat to a fishing grounds northwest of Kirkjuvogur:

> When we had just got to the fishing grounds . . . we all heard an uncanny thunderous roar southwest in the sea. What I call a roar happened in a curious way, for the boat we were in juddered and creaked as if all its nails would shake free, and it seemed to us that the sea too was agitated, which could not, however, be ascribed to a storm, for the weather was calm. That screeching or strange noise also had the effect that most of the crew were beside themselves, and felt that bones were scraping at their skin, and their muscles and sinews trembled.[6]

In his *Gare-Fowl Books*, Wolley makes only scant reference to such supernatural perils at sea. He may have dismissed such tales, or perhaps the witnesses were reluctant to share them with outsiders?

"Extraordinary to See"

On the latest successful trip, Hákonarson determined the course of the boat as he went along, consulting with other experienced members of the crew. Probably he paid particular attention to the eldest crewman. As boat foreman, Hákonarson had to ensure that the oarsmen rowed in time, keeping up a steady speed; he would try to avoid breakers, navigate through troughs, and aim to reach their destination by the shortest route. Fortunately, the weather stayed favorable. They rowed out of the bay at Kirkjuvogur, then turned southward along the Reykjanes peninsula, rowing onward until they reached the "toe," where they headed southwest out into the open ocean. It was a long, light spring night; at that time of year in Iceland, the sky never gets completely dark.

Crossing the Reykjanes Race, of which Eldey is the nearest island, was the most dangerous part of their voyage. It was often fatal to travelers in open boats. A vortex sometimes forms there, off the southwestern tip of Iceland, where two ocean currents meet, one from the southwest, the other out of the north-northwest. It was essential to advance with care, navigating between the waves where the two currents met. Shortly before reaching Eldey, Hákonarson's boat encountered a French schooner. The French, astonished to see men rowing an open boat in such strong currents, indicated they were welcome to come aboard, but Hákonarson ignored them. He knew what he was doing.

At about six in the morning, after twelve hours at sea, Hákonarson and his men reached Eldey, breathless and exhausted. The morning sun was shining as they steered in close to the perpendicular cliffs; the sea boiled at their foot, as if a volcanic eruption were in progress beneath the rock. Gannets and fulmars formed horizontal white stripes on the cliffs as they sat neatly on their nests—as if on the balconies of a natural theater. Everywhere there were birds: flying around the rock in search of food, diving down into the sea around the boat, or returning to their nests.

The boat reached the ledge where bird-hunters had sought their prey in recent years, the "lowland" or "underland," as the seamen called it. Hákonarson spotted two great auks, in among clusters of other, smaller birds. He and his crewmen knew that great auks tended to stay in a fissure in the ledge, and were best spotted from out at sea. In Ketill Ketilsson's account, according to Wolley: "They saw the two birds as they approached the incline

FIGURE 8.4. Family shield with four birds. Drawing by Alfred
Newton during an interview. John Wolley's *Gare-Fowl Books*, 1858.
(Reproduced by kind permission of the Syndics of Cambridge
University Library.)

beneath the cliff. Their white breasts were turned out toward the
ocean." It seemed as if the two great auks were hosts awaiting
visitors; they welcomed them with dignity, upright and stately.

Wolley and Newton were no doubt impressed by this description. As with some of their own sketches, it seemed to

PLATE 1. Ornithologist Jan Bolding Kristensen, Natural History Museum, Copenhagen, attending to a "fake" great auk. (Photo: Gísli Pálsson.)

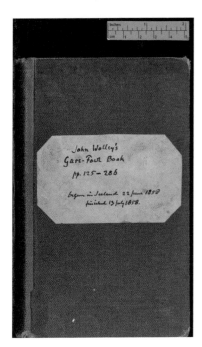

PLATE 2. John Wolley's *Gare-Fowl Books*, 1858. Book 2. (Reproduced by kind permission of the Syndics of Cambridge University Library.)

PLATE 3. A great auk egg (nr. 4832) belonging to John Wolley. Drawings from *Ootheca Wolleyana*, artist Henrik Grönvold (1858–1940). The egg was captured on Eldey, Iceland, in 1835. (Photo: Gísli Pálsson.)

PLATE 4. John Wolley. (Courtesy of Balfour and Newton Libraries, Cambridge University.)

PLATE 5. Pepys Library, Magdalene College, Cambridge. (Photo: Gísli Pálsson.)

PLATE 6. Painting by Nicolas Robert, sometime between 1666 and 1670. (Courtesy of the National Library of Austria.)

PLATE 7. Painting by Aron from Kangeq, 1868. (Courtesy of the National Museum of Greenland, Nuuk.)

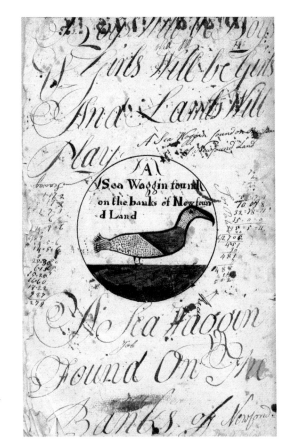

PLATE 8. "Sea-Woggin at Newfoundland." Drawing by Abraham Russel, 1793. (Courtesy of the New Bedford Whaling Museum.)

PLATE 9. Eldey, or the "Meal Sack." The "underland" of great auk nesting to the right. Snæfellsnes Glacier in the distance. (Photo: Ragnar Th. Sigurðsson.)

PLATE 10. "A Rough Road—Iceland." Drawing by Bayard Taylor, 1862. (Library of Congress, Washington, DC. Public domain.)

PLATE 11. "Awkward Predicament, Icelandic Custom." Drawing by Bayard Taylor, 1862. (Library of Congress, Washington, DC. Public domain.)

PLATE 12. The skerry Karl by Reykjanes, Eldey in the distance. (Photo: Gísli Pálsson.)

PLATE 13. Sigríður Þorláksdóttir stuffing great auks, 1831. Scenery by George Dawson Rowley and his wife, Caroline, four decades after the event. (Photo: Errol Fuller.)

PLATE 14. *Great Auks in a Mist*. Painting by Errol Fuller. (Photo: Errol Fuller.)

PLATE 15. Errol Fuller attends to a great auk egg at his home. (Photo: Gísli Pálsson.)

PLATE 16. Alfred Newton, 1866. Painting by C. W. Furse. (Courtesy of Balfour and Newton Libraries, Cambridge University.)

highlight the grandeur and proudness of the bird. Nature-lovers of the Victorian age often saw the natural world as reflecting the laws of human society, with rich and poor, high and low.[7] Did the two British gentlemen imagine that these stately great auks would have been right at home amid the elite of Eton College?

The Hunt

Great auks were in their element in the sea. There, the birds could easily swim away from hunters and, if necessary, could dive, surfacing a long distance away. One time when a local foreman led a group of men to search for birds on the cliff, they spotted six birds but managed to grab only one by its tail feathers: the rest jumped off the cliff, from a height of sixty feet, and were not seen again. But on the rock, a great auk was helpless: a "sitting auk." One witness told Wolley and Newton he had once caught eight birds. He broke their necks; he had no time to wring them.

Now, on the latest successful trip, a strong undertow tossed the boat about by the rock, although the weather remained fine. High waves washed the foot of the cliffs, and it was necessary to take care landing, to avoid ramming into the rock and damaging the boat or losing a man overboard. As Hákonarson well knew, foremen had sometimes lost crew members at the island. Speed, for the hunters, was of the essence. The men had to move fast, and get the boat away from the rock as quickly as possible. Hákonarson wanted four men to go up onto the ledge to catch the birds, but one of them refused to leave the boat, unwilling to take the risk. The other three—Jón Brandsson (1804–80), Ketill Ketilsson (1823–1902), and Sigurður Ísleifsson (1821–89)—rushed ashore.

Wolley retells the full accounts of several of the men who went on the trip, including Ketilsson.[8] He and Ísleifsson ran up

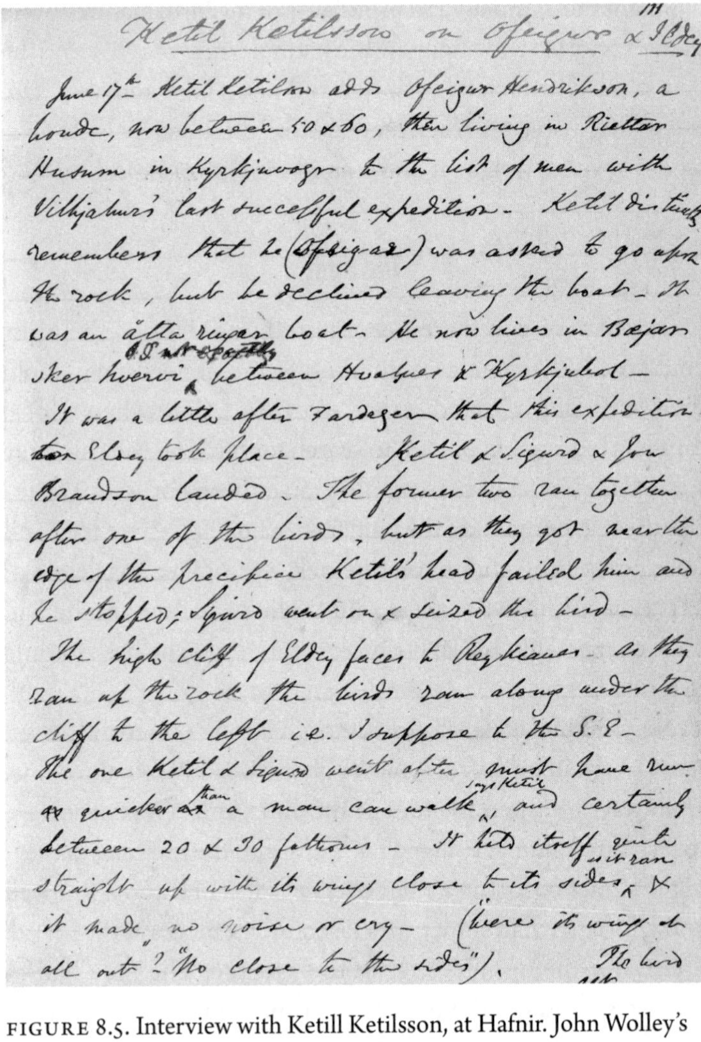
111

Ketil Ketilsson on Ofeigur & Eldey

June 17th – Ketil Ketilsson adds Ofeigur Hendriksson, a bonde, now between 50 & 60, then living in Riettar Husum in Kyrkjuvogr to the list of men with Vilhjalmur's last successful expedition – Ketil distinctly remembers that he (Ofeigur) was asked to go up the rock, but he declined leaving the boat – It was an ätta rigar boat – He now lives in Bæjarsker hverri, I.S. not exactly, between Hvalsnes & Kyrkjuból –

It was a little after Fardagar that this expedition to Eldey took place – Ketil & Sigurd & Jon Brandson landed – The former two ran together after one of the birds, but as they got near the edge of the precipice Ketil's head failed him and he stopped; Sigurd went on & seized the bird –

The high cliff of Eldey faces to Reykianes – As they ran up the rock the birds ran along under the cliff to the left i.e. I suppose to the S.E. – The one Ketil & Sigurd went after must have run as quicker than a man can walk, says Ketil, and certainly between 20 & 30 fathoms – It held itself quite straight up with its wings close to its sides as it ran & it made no noise or cry – (were its wings at all out? "No close to the sides"). The bird

FIGURE 8.5. Interview with Ketill Ketilsson, at Hafnir. John Wolley's *Gare-Fowl Books*, 1858. (Reproduced by kind permission of the Syndics of Cambridge University Library.)

the ledge after one of the birds, but when they came to a vertical cliff, "Ketill's head failed him and he stopped." Ísleifsson went on and caught the bird, remarked Ketill. "It must have run faster than a man—it ran . . . upright, with its wings by its sides, and gave no sound or cry." But why did Ketilsson's head "fail him," as Wolley writes (presumably as translated by Zoëga)? Was he suddenly dizzy, a coward, or was he responding as a moral being unable to do the deed when the moment of truth had come?

Wolley didn't ask. Instead, he interrupted the narrative to ask whether the bird had flapped its wings as it fled. No, replied Ketilsson. The wings lay close to its side. The second bird, chased by Jón Brandsson, did not go far. Ketilsson thought the bird had been incubating an egg. Wolley reports that his witness, Ketilsson, who had apparently found the chase alarming, went to the bird's nesting site on the rock, where he found a black-speckled egg. "It was cracked or broken on the side upon which it was lying, i.e., the side next the rock. Ketilsson laid it down again where he had found it." This statement has been quoted and retold—and embellished—numerous times.

Jón Gunnarsson, who had remained in the boat while the three men chased the birds, sat listening to Ketilsson's account, in Wolley's presence. He apparently made no objections to Ketilsson's testimony. Sigurður Ísleifsson then gave his account: Jón Brandsson had hurried up across the "lowlands" with his arms spread wide. The bird tried to get away, but Brandsson caught it. Ísleifsson says that he then headed for the rock, and caught the other bird on the edge of the precipice, with the sea directly below him. Wolley writes: "They were close under the perpendicular cliff. His bird ran certainly from where Sigurður first saw it, about twenty fathoms [one hundred and twenty feet]. It walked like a man . . . but moved its feet quickly. It began to run as . . . the razorbills. . . . Its wings lay close to its side, not hanging out (on being asked these last two questions). As he held it by the neck it hung out with its wings a little (volunteered). . . . Saw nothing of eggs."[9]

Hákonarson also gave Wolley an account of an expedition during which two birds were caught. Interestingly, this trip may not be the same one that Ísleifsson and Ketilsson described, complicating the narrative and calling into question the list of crewmen so painstakingly established by Wolley and Newton; Wolley was not certain. On this "latest trip," there were fourteen men on two boats, Hákonarson said: Jón Brandsson, Ketill Ketilsson, and Frímann Gíslason (who is not included on Wolley and Newton's list) went up onto the ledge. Gíslason led the way, with a rope around his waist and in his stockinged feet, to get a better grip on the slippery surface. The three men quickly caught both birds and came straight back to the boat, as the sea was rough. They saw two eggs, according to Hákonarson, but left them behind. Gíslason and Ketilsson boarded the boat without difficulty, but Brandsson was fearful. "They told him to tie a rope around himself and throw himself in the sea, but he would not do. He had rather stay where he was. So Hákonarson said 'If not I will take you with the boathook'; to which Brandsson tied himself and was put through the sea. It was such Satan's weather that they had not time to pick up the eggs."[10] So Brandsson was dragged aboard.

On the "latest successful trip" that Ísleifsson and Ketilsson described, the rest of the crew watched in anticipation as the birds were chased and caught. Guðni Hákonarson recalled that the birds had run up the slope when they heard the men coming, as that was the most direct way for them to go up the ledge and fling themselves into the sea. The birds made no sound, said Guðni—neither on that occasion, nor on previous trips to Eldey. Guðni told Wolley: "When one comes up the rock the other birds begin to cry and the geir-fowl comes down to meet one, comes walking slowly (Guðni makes action) like children, but quite upright, neck straight up. Guðni speaks this last with emphasis, decidedly as though he remembers well."

The wind was rising and the boat swung on the waves. The men who had caught the birds hurried to the edge of the rock, passed the birds down to Vilhjálmur Hákonarson, released the rope and jumped aboard—on this trip, no one was dragged through the sea—and Hákonarson instantly gave the order to row away. The crew were tired by their exertions, so their homeward journey took even longer than the trip out to Eldey; the boat was not beached until evening.

The exhausted oarsmen dispersed to their homes, most in the hamlet surrounding Kirkjuvogur, where they were given a warm welcome. Mothers and wives had been waiting anxiously. The following day, Hákonarson set off to Reykjavík to sell the two birds. Along the way, he met by chance the speculator Christian Hansen, who was then living on his vessel near Keflavík. Hansen offered him a better price than he could expect from Carl F. Siemsen, the Reykjavík merchant who had originally urged Hákonarson to make the trip. Hákonarson had made no binding agreement with Siemsen, so he accepted Hansen's offer: eighty *rigsdaler*, equivalent at that time to nine pounds sterling (worth approximately $750 today).

On his return home, Hákonarson settled up with his oarsmen and paid them their share of the profits. One was paid in cash ("five specie *daler*"—that is, silver coins), plus coffee and sugar. Another appears to have received a cask of rye meal; perhaps Wolley found that amusing, given the English name of the "Meal Sack." Some of the farmhands were paid in butter and meat. Once the winter fishing season and the great auk hunting expedition were over, some of the men rode north, as usual, to work elsewhere for the haymaking season, carrying their goods with them on pack-horses that they had brought south the year before.

Christian Hansen passed the two birds on to a pharmacist in Reykjavík, who commissioned the French artist Vivien to paint a picture of one of them; that was the unattractive or inaccurate painting Wolley and Newton had seen before they left Reykjavík. After that, the birds embarked upon their journey out into the wider world. Wolley quotes a merchant as saying that the birds' eyes and viscera had been preserved in alcohol in Copenhagen, and were greatly valued. Asked by Wolley whether he was sure that both birds were there, he said no, perhaps only one of them. These are probably the great auk organs exhibited in large jars at the Natural History Museum in Copenhagen, but it's not certain. For a long time, the whereabouts of the skins of these historic great auks remained a mystery. Recently, forensic scientists have located the male's skin in Brussels; the female's is still at large.

After hearing these tales of the "latest successful trip" to Eldey, Wolley and Newton were still hopeful that the great auk had not permanently disappeared. Because more than one egg was said to have been seen, in some accounts, they deduced that another pair of great auk could have escaped to sea when the two birds were taken. They took consolation in the fact that the great auk did not nest every year on Eldey; sometimes several years would pass before they turned up again. So they could still hope to hear of great auks elsewhere.

A Far Simpler Tale

Readers of the *Gare-Fowl Books* may be surprised that nothing in the locals' accounts suggests that people intended their hunting to be kept secret. In the subsistence economy of

Icelandic coastal communities, there is a long-standing tradition of catching seabirds and collecting birds' eggs; these were an important source of nutrition. The hunting of the great auk was only different in that, from the mid-eighteenth century on, it was impelled by new market forces: the growing demand from collectors and museums abroad for the birds' skins and eggshells.

The *Gare-Fowl Books* reveal that Wolley and Newton often had difficulty grasping the facts based upon differing verbal testimonies. Sometimes they added a correction some days after an interview, when they felt they understood things better. Occasionally, they reached a judgment based on the reliability of an account, or rather of the person giving it; the narratives were contradictory, or unconvincing, and some of the local men were more interesting than others. The two British naturalists felt the need to simplify this complicated story: to interpret their inconsistent sources and cook up an easily digested version of the truth for their contemporaries, as well as for future generations.

Newton indicates that he had no way of reconciling all the inconsistencies, and that it was necessary to tell a far simpler tale. He states some provisos, defending his position:

> I do not claim entire novelty for several of the statements I have to make. . . . Nor do I profess to be sure that the account I have to give is always the true one. It must be remembered that the results here recorded are the main points of evidence deduced from many authorities . . . and though I do not doubt that the greater number of these latter are persons of eminently truthful habit (for such is the natural characteristic of the Icelander), yet some few there are who may have willfully told falsehood.[11]

Advocates of Victorian science on a quest for the truth about
the relationship between humans and the great auk, Wolley and
Newton probably gave some thought to where to draw the line
between fact and fable, history and myth. But they never mentioned the issue explicitly, and deemed some evidence credible,
although documented centuries ago.

Their doubts about the reliability of their informants, and
the apparent or evident inconsistencies in the narrative they
documented, are not surprising to the modern fieldworker.
Such doubts are endemic in anthropological works. Sometimes surreal myths—accounts of elves or howling underwater
beings—are "understood" as being coded language or metaphor.
Sometimes they are taken quite seriously, as references to the
hidden, incomprehensible forces of nature. The experience of
fieldwork is a complex terrain, necessarily informed by social
context and complicated by a host of constraints, ranging from
the mood of the fieldworker to the cultural distance between
the observer and the observed (in colonial language), and necessarily biased by social identity (gender, race, and social class,
for instance), as well as by prior experience. While the problem
of interpretation is sometimes exaggerated by poor language
skills and cultural differences, it is practically everywhere in
anthropological research.

It is interesting to observe that when Wolley and Newton
state the ages of the crew members on the "latest successful
trip," they calculate them based on the year 1844. Their witnesses, however, as we have seen, were not all in agreement on
the date of this apparently momentous event. Hákonarson said
that he had "again thought over the date of the last visit to
Eldey when they got birds, the two birds, he is sure it was 1849
because it was five years afterward that he went again, and now
it is four years since then." A farmhand appears to confirm

Hákonarson's testimony, but the accounts given by these two men are inconsistent with much that the other members of the crew maintain.

It is understandable that the hunters were inconsistent in their accounts, bearing in mind that they had no inkling that an expedition they had made a decade or so earlier was in any way remarkable. The calendar of Icelandic rural society functioned according to the changing seasons, and not any official chronology. The hunters did not have the technology to measure time precisely, although they observed the heavenly bodies and the weather and took account of wisdom handed down by generations. Calendar dates were not that important to them. Some, as Wolley noted, did not even keep track of how old they were.

Wolley and Newton speculate that the discrepancies in the stories told by the local men can be attributed to fallible memory. In 1861, Newton put forward a clear version of the tale in a brief essay in the periodical *Ibis*, which was widely disseminated. Writing only three years after he returned from Iceland, he says:

> The last Gare-fowls known to have occurred in Iceland were two in number, caught and killed in 1844 by a party, of which our excellent host at Kirkjuvogur, Vilhjálmur Hákonarson, was leader. . . . As many persons may regard these birds as the latest survivors of their species, I may perhaps be excused for relating at some length the particulars of their capture, the more so as this will serve to explain the manner followed on former occasions.[12]

Newton's essay is written with great care and attention to detail. *Ibis*, the International Journal of Avian Science, was and remains an essential publication for ornithologists. Newton was one of the founders of the journal, and edited it himself for a time. His 1861 essay is one of the major sources cited regarding the last of the great auks. In time, his account took on a life of its own—like the eggs and bird-skins in collections around the world.

In his book *Who Killed the Great Auk?* Jeremy Gaskell argues that the 1849 expedition was indeed made, but that another trip had been made, with similar results, possibly five years earlier. This could explain the apparent contradictions in the evidence of the witnesses.[13] The third volume of the *Gare-Fowl Books* contains a summary of great auk hunting expeditions from 1828 onward, including Hákonarson's "latest trip," when he caught the two birds. Significantly, Wolley the scribe is uncertain about the dates, repeatedly correcting himself as he reported expeditions from 1842 to 1849.

Gaskell's arguments are persuasive; yet it is curious that Hákonarson, who was living cheek-by-jowl with the two British naturalists for weeks on end, should not have distinguished between two successful hunting trips he had made, five years apart. Wolley remarks:

> Though Vilhjálmur cannot quite provide the clearest proofs of each of his opinions, it is evident from the care with which he speaks with the almanac before him, that he speaks to the best of his ability, and that that knowledge of his is founded on a consideration of tradition and of all the facts that have come to his notice, all which he as the leader of the later parties has been obliged to keep in mind.[14]

Wolley had confidence in his witness. But what then had happened to the two birds caught in 1849—if Hákonarson's account and Gaskell's interpretation are to be relied on? It is impossible to tell.

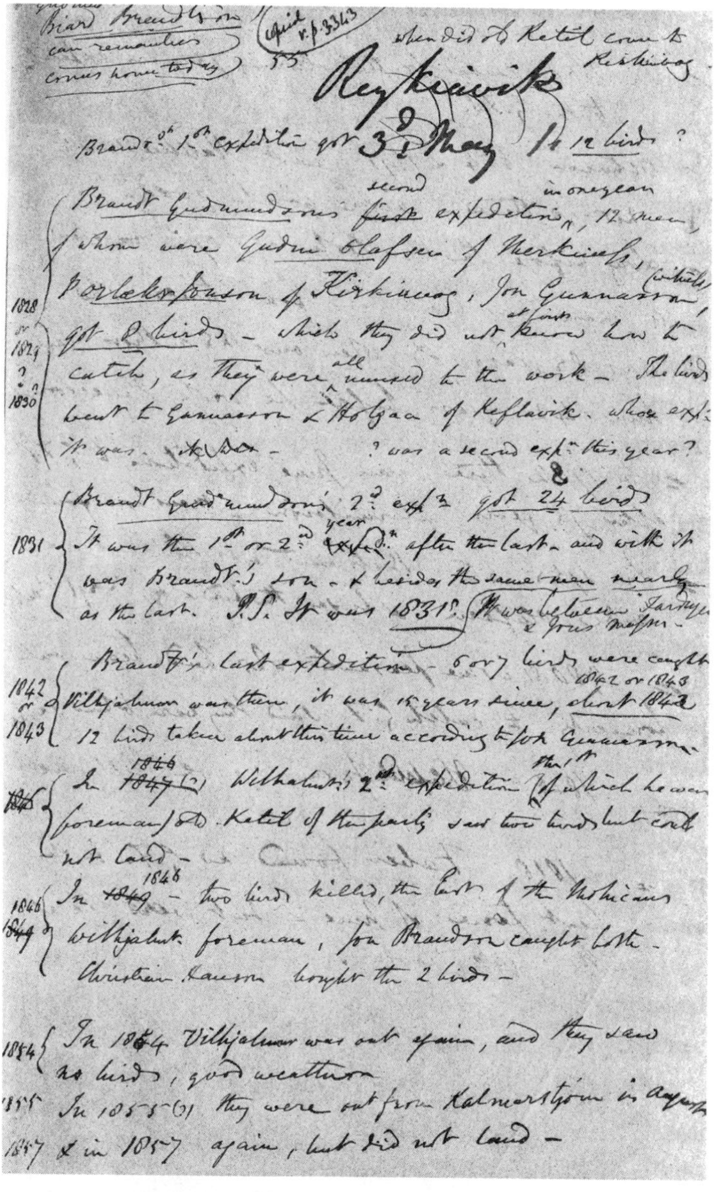

FIGURE 8.6. Great auk expeditions, listed by John Wolley and Alfred Newton. John Wolley's *Gare-Fowl Books*, 1858. (Reproduced by kind permission of the Syndics of Cambridge University Library.)

Translation Problems: Last or Latest?

Occasionally, a native speaker of Icelandic will sense underlying problems of translation while reading the *Gare-Fowl Books*, although these are not spelled out by Wolley or Newton. The guide and translator they had hired, Geir Zoëga, was experienced, and overall he seems to have done a reasonably good job. And, given that Wolley and Newton had some understanding of the Icelandic language, gross errors of translation are unlikely. But the text rarely mentions potential debates about nuances in interpretation of some of the statements made by the hunters, simply avoiding several important issues that today we would like to see fleshed out more fully.

One case is the frequent mentioning of the 1844 trip, as "the last" or "the latest" one (presumably *síðasti* or *seinasti* in the Icelandic original). Did interpreter Zoëga fail to make the distinction or draw attention to it? Was it Wolley who jumped to conclusions? In any case, he opted for the dramatic "last." Another example is the statement that Ketill's "head failed him" on the rocks. The text doesn't explain whether he was feeling dizzy or unable to carry out the deed; much depends, again, on how this is read—keeping in mind twentieth-century debates about the ethics of killing endlings, the last exemplars of a species. Trivial as these examples may seem, they represent some of the key statements of the *Gare-Fowl Books*. Unfortunately, Wolley did not elaborate, unaware of any historical repercussions of his narrative for the unfolding of the discourse about extinction in Britain and beyond.

The modern reader of Wolley's manuscript is struck most of all by a peculiar silence: there is not a single reference to "extinction" in the entire corpus. It occurs, however, at the end of Newton's 1861 essay about the expedition in *Ibis*, in the context of the role of human impact, where he writes, "Whether the Gare-fowl be already extirpated or still existing in some unknown spot, it is clear that its extinction, if not already accomplished, must speedily follow on its rediscovery."[15] While extinction seems to hover like a heavy and mystic cloud above both examiners and the examined, to use Wolley's terms, the language is simply not in the *Gare-Fowl Books*. The great auk, the key subject of two months of intensive discussions, was just "getting very rare," Wolley wrote, and possibly being forced to move from one nesting place to another. The task remained for Newton to make sense of their experience and to make his conclusions heard among scientists and environmentalists in Britain at a time of swift conceptual and political upheaval, partly focused on ideas of speciation and extinction.

CHAPTER 9

THE HUMAN DRAMA

Long after the days of Wolley and Newton, British writer and artist Errol Fuller followed in their footsteps with his magnum opus, *The Great Auk* (1999), a fine work of reference drawing upon his extensive collection of great auk memorabilia.[1] The Victorian age is supposed to be long over, yet the threads that were spun in that era remain strong, and one of them winds through Fuller's house in Tunbridge Wells, just south of London. Fuller has collected endless documents and objects; he has made his home into a cabinet of curiosities itself. On each of its three floors is a jumble of stuffed birds—rare or extinct—birds' eggs, stuffed monkeys and fish, antiquities, glass cabinets, musical instruments, fossils, sculptures, animal bones, books, manuscripts, documents, and paintings (see plate 14). One glass cabinet contains dozens of stuffed hummingbirds, neatly mounted on tree-branches. On a bathroom wall are two gigantic lobsters in glass cases, as if they had clambered up out of the bathtub. Among the most interesting items connected to the great auk are two large, gleaming hardwood cabinets labeled "Newton collection, Ootheca Wolleyana, no. 1." Fuller says he bought the cabinets cheaply: the museum in Cambridge saw no reason to keep them.

FIGURE 9.1. Errol Fuller and his cabinet of curiosities. The cupboard was part of John Wolley's egg collection. (Photo: Gísli Pálsson.)

Errol Fuller was born in Blackpool in the north of England, and grew up in south London. His father was a police officer and his mother a housewife. They lived in social housing, without any luxuries. When he was seven or eight, he remembers, his mother started to take him with her when she went out, as she could not leave him at home alone. Early in the morning, they would take a train to Charing Cross Station in central London. Fuller's mother would leave him at the National Gallery or the British Museum, where he would have safe refuge while she did her shopping, and he would not slow her down. She probably hoped that he would learn something, too. So the museum and gallery curators became, all unknowingly, child-minders for the policeman's wife. Mrs. Fuller urged her son to look around the collections, but told him that he must be sure to meet her under the clock in the lobby at four o'clock precisely, when they would go home. Fuller was always ready on the dot. But he found plenty to interest him as he whiled away the day.

He roamed the galleries, stopping here and there, captivated by some historical natural object or work of art. In his home district in south London, birdlife was minimal—mostly confined to the odd pigeon. But he was not primarily interested in live birds. It was stuffed birds that caught his imagination, drawing his attention to rare species, and extinction—subjects that have remained with him ever since. Leaving school after completing his compulsory education, Fuller channeled his energies into his passions: painting, birds, all sorts of natural objects, and boxing. He taught himself what he needed to know, sometimes under the guidance of helpful people who knew more than he did.

His interest in the great auk began when he was a little boy: his grandfather showed him a bird book, which included a couple of pages about the great auk, and so the bird slipped into the boy's consciousness. All his life, he remarks, he has been gathering material about the great auk. And now he has his *Wunderkammer* in Tunbridge Wells, and is a respected author, painter, and ornithologist—one of the world's foremost experts on extinct bird species.[2] One of his books, *Drawn from Paradise*, which he wrote with David Attenborough, has sold forty thousand copies.[3]

Fuller's collection includes two great auk eggs. Their packaging is formal, in the Victorian manner, but would hardly meet the standards of a modern natural history museum. One egg is in a drawer in a small chest. Fuller pulls out the drawer and proudly shows off the large egg without touching it. The room is in semi-darkness, but a ray of bright light seems to fall on the egg when the drawer is opened, like a spotlight that draws attention to the treasure and casts a mystical glow upon it (see plate 15).

Fuller had long cherished a desire to acquire a taxidermied great auk, but the owners of such birds guard them with their lives, and when they are offered for sale, the price is high. About a decade ago, he did acquire a great auk from Eldey island—or, more precisely, a share of the bird, which he bought with a friend. "That was a mistake," he remarks. Fuller kept the bird for a while, and enjoyed having it around and looking at it. But then his co-owner said: "Now it's my turn," so Fuller had to hand it over. Perhaps both of them realized that the vulnerable stubby-winged bird could not continue migrating back and forth. At any rate, Fuller's co-owner received a generous offer for the bird and concluded the sale, without consulting him, although he sent him his share of the profits.

It transpired that the new owner of the bird was a former education minister of Qatar, Sheikh Saud Al-Thani, who became Fuller's friend. Al-Thani died shortly afterward, and the great auk was placed in the collection of a museum in Qatar. That bird is one of the twenty-four great auks that were brought ashore in the spring of 1833 by Brandur Guðmundsson and his men, and was taken to Sigríður Þorláksdóttir of Njarðvík, who "was out of doors all day in the fine weather, stuffing birds" that spring, as reported in Wolley's *Gare-Fowl Books*. Flightless though it was, this great auk has traveled far.

"Short of Nothing"

It was mid-July 1858. John Wolley and Alfred Newton had spent eight weeks at Kirkjuvogur, but now new tasks awaited them back home in Britain. Even if it had nested that year on Eldey, the great auk by then must have vanished out into the vast expanses of the ocean; the local people, too, had their summer work to do. Wolley and Newton packed up their possessions,

FIGURE 9.2. Errol Fuller's great auk, now in Qatar. (Photo: Errol Fuller.)

the *Gare-Fowl Books*, and some bird-skins, and made a hasty excursion to the Geysir geothermal springs—like true modern-day tourists—before they left Iceland. British artist and activist William Morris (1834–96), who rode to Geysir later, in 1873, hoping to become acquainted with Iceland in peace and quiet, discovered that it was already a popular tourist attraction. When he arrived, it was "all bestrewn with feathers and wings of birds, polished mutton bones, and above all pieces of paper."[4]

Wolley and Newton sailed home via the Faroes. Back in England, Newton visited his childhood home at Elveden, then went on to Cambridge. The two men had learned a good deal in Iceland, yet both were sorely disappointed. In a letter written at Elveden on 16 August 1858, Newton wrote:

> The result, then, is short of nothing. Not one day of the whole two months we were at Kirkjuvogur was the sea ever sufficiently calm to have allowed us to land, even had we gone out, and we have come back knowing no more than when we started whether the Great Auk is living or dead.[5]

Wolley was the driving force behind the Icelandic expedition; he was the adventurer and collector, he handled most of the logistics, and the *Gare-Fowl Books* are mostly his work. Newton's scholarly status, on the other hand, lent the project academic heft. No doubt both men foresaw this journey as increasing their renown among naturalists and scholars in Britain and elsewhere, as both were in a precarious position, without tenure, unmarried, and dependent upon others. The two men were clearly close friends: they "walked in step." During their long and demanding expedition in Iceland, they were inseparable, respecting each other's interests and dividing tasks between them; they were equal in everything. There is no indication that one man sought to take priority over the other or wanted more credit. The great auk expedition was a collaborative venture that could be expected to yield benefits for both; it was possibly the high point of their careers.

Now, having returned from Iceland with so little to show for their efforts, the two friends found themselves having to reboot their expectations once again. Newton had his eye on a secure position at Cambridge, a prestigious post that would enable him to pursue his research on birds and bird protection for years to come. He felt that the time had come to write a comprehensive study of the great auk. Wolley probably foresaw the end of his egg-collecting activities—sooner or later, that time-consuming and eccentric pursuit must come to an end. But he planned to catalogue his vast collection, record a detailed account of it, and ensure its future home.

As the year passed, their connection with Iceland was not broken, in spite of the distance. Shortly after his return to Britain, Wolley sent a letter and gifts to Director of Public Health Jón Hjaltalín as a token of gratitude for the help he had provided in Iceland. Hjaltalín replied immediately with his thanks, and alluded to a possible further visit by Wolley:

> I wish very much that we might have the pleasure of seeing you next year; you must not be afraid for our rude climate, it is not always so bad as it was during the last summer. You will most likely succeed on your next voyage and then what a satisfaction would that be for you and your worthy friend Newton [to] succeed in your very interesting and scientific investigations.[6]

Wolley and Newton would never return to Iceland, but they appear to have urged Geir Zoëga to look out for a great auk the next spring, and to bring it to England if he was passing that way.

Zoëga had been responsible for packing up the eggs and other items collected by Wolley and Newton and placing them aboard ship. In the summer of 1859, Zoëga wrote to Wolley:

> I am sorry to write to you, that your anxiety, about arriving to England with Geirfugl is not likely to be realized this summer. . . . I will, if anytime I shall be successful to catch one or several of the birds, do my best in complying with them, and take care of them (or it) as tenderly as of the pupils of my eyes. . . . Now at last I have only to thank you for your kindness . . . last summer. I wish very much to come to England, particularly bringing at last a living Geir-fugl with me. Nothing in my mind shall be omitted for this purpose, and I am sure that the same is the case with Vilhjál-mur Hákonarson.[7]

Newton stayed in touch with Hákonarson until the latter's death. In his essay about the Icelandic expedition, he writes that two years after the expedition, on 13 June 1860, Hákonarson succeeded in landing again on Eldey, but found no great auk.[8] Newton owed a debt of gratitude to Hákonarson, and freely acknowledged it. At the end of his essay, he expresses his thanks to the people of Iceland for their friendly reception: "In Iceland all, with but one exception, were eager to tell us all they knew, and that in the most careful manner."[9] (It is not clear who the sole exception was.)

In 1870, Hákonarson's daughter Anna eloped with theologian Oddur V. Gíslason (1836–1911), and a year later "Vilhjálmur the Wealthy" died from melancholia. His daughter's rebellious marriage was said to have killed him: "The anguish suffered by the father for this cause affected him deeply, to such a point that it is believed to have led to his grave illness and suffering."[10]

The Sad Fate of John Wolley

Wolley and Newton hardly discussed their private and emotional lives, even though they had lived cheek by jowl in Iceland. They corresponded regularly after the end of their Iceland expedition, but their letters are generally objective, distant, and impersonal—like the *Gare-Fowl Books*. Newton never married, and his attitude toward women was complicated. While he was invariably polite, he dismissed out of hand any suggestion that women might be admitted to his inner sanctums to peruse his collections, such as the eggs he held so dear. He was outraged by the proposal that women might be admitted to the chapel at Magdalene College, where he lived and worked (women did not gain admission until after his time).[11]

Wolley's attitude to women was probably somewhat different from that of Newton. On his arrival in Britain after the Iceland expedition, he hoped to establish a family. One of his personal letters to Newton is particularly noteworthy. On 3 September 1858, shortly after his return, Wolley wrote with unusual candor about his future plans and his financial situation. His letter commences, as usual, with some words about great auks, then goes on to more personal matters:

> But I have important news to tell you about myself, not favourable to future ornithological exploits. I am engaged to be married as soon as I can get something to do to enable me to keep a wife—you will groan at me, but forgive me, and tell me how can I get a decent living for two. . . . What is the new museum at Oxford? If you have any advice to give me . . . I shall be very grateful. My future wife's name is Janetta Loraine. . . . She has been making a long stay among our relatives with my sister Ms. Ricketti her most intimate friend. Of

course you must not expect to see even an "almost goodlook-
ing" person, but one who has unlimited goodness of heart,
and many qualities that suit her to be a poor man's wife. She
is twenty-two years of age—for the rest you must wait and
use your own experience.[12]

Wolley's fiancée, Janetta Hannah Loraine, was born in Fawdon,
a suburb of Newcastle, in 1836. She was an orphan, the eldest of
four children of John Loraine and his second wife, Caroline Isa-
bella (née Ekins).

Wolley's unprecedented confidence about his engagement and
marriage, and the limitations his new lifestyle might mean for his
ornithological future, can be read in various ways. But Wolley
clearly realized that it would entail some kind of parting of the
ways with Newton: at the very least, his marriage would disrupt
their intense collaborative relationship. Two days later, Newton
thanked Wolley for his letter, offering his congratulations.

Wolley's last letter to Newton was written a year and a half
after they returned from Iceland. During that period, the two
friends exchanged dozens of letters, and much happened in the
life of each. But as time went on, Wolley's letters became harder
and harder to decipher, even to one well accustomed to his
handwriting. Not only is the writing unclear, but also the con-
tent is erratic. Wolley's friend must have realized that his memory and concentration were failing.[13]

In fact, soon after announcing his engagement, Wolley fell ill
and had to abandon all his future plans for marriage, employ-
ment, and renown. The evidence indicates an ailment in the
brain, possibly due to a viral infection or tumor; the symptoms
included failing memory and loss of presence. Wolley had always
been physically fit and had undertaken many challenges as he
sought out birds in remote and difficult habitats, but now he lost

all his strength and vigor. Sometimes his condition seemed to improve, but after a traumatic experience in July 1859, his doctor informed him that he did not have long to live. Wolley recounts the experience in a letter to Alfred Newton's brother, Edward:

> I had a sudden fright from my father being nearly run over on a railway, a steep incline. In walking behind this frightened me terribly although he saw the train in time to make him yield to me, and he missed being run over by most narrowly. The consequence of this fright has been a serious illness which alarmed my friends a good deal. It was of course my previous weak nervous condition that made me so easily act on.[14]

John Wolley died on 20 November 1859, aged thirty-six. The person most affected, apart from Wolley himself, was his fiancée, Miss Loraine, whose future was thrown into confusion. No other evidence of their relationship is preserved in the library in Cambridge where Wolley's correspondence is kept. According to the 1861 UK census, she was a scholar, living in St. Margarets, London. Genealogical sources on the web reveal that she eventually married Richard Barrow, a widower. The couple had one child, Claude Loraine Barrow, who was born near Matlock, on 10 August 1870. Tragically, Janetta Hannah died eighteen days after his birth. Her son attended Jesus College in Cambridge and became part of a scientific expedition surveying the alps in New Zealand.[15]

"Abstract of Mr. J. Wolley's Researches"

Wolley bequeathed his egg collection and all his documents to his friend Alfred Newton. It fell to Newton to write a report on their Icelandic expedition and to ensure a secure and permanent home for Wolley's eggs and archive.

Newton wrote an extensive obituary of Wolley in the ornithological journal *Ibis*, in which he bids his friend a sad farewell.[16] In honor of Wolley, and as a token of his gratitude and a testimony to their profound friendship, Newton put aside his plans to write at length about the great auk and instead commenced work on his magnum opus, a description of Wolley's egg collection. The work would ultimately comprise four volumes: *Ootheca Wolleyana: An Illustrated Catalogue of the Collection of Birds' Eggs Begun by John Wolley* was published between 1864 and 1907.[17] It begins with a long memorial tribute to Wolley; Newton thus bids farewell to his friend once again.

Ootheca Wolleyana would take up much of Newton's time for the rest of his life; the last volume was published in the year of his death. This complex and time-consuming work was more important for Newton than many modern readers might think. It was a delicately illustrated atlas somewhat in the fashion of the time, serving, as Daston and Galison put it in their seminal work *Objectivity*, "to train the eye of the novice and calibrate that of the old hand. [Atlases] teach how to see the essential and overlook the incidental, which objects are typical and which are anomalous, what the range and limits of variability in nature are."[18] Together with Audubon's glossy work *Birds of America*, the monumental *Ootheca Wolleyana* offered a professional gaze into the avian world. One wonders if the scientific atlases might have something in common with the stunning artistic representations of birds and other animals on Paleolithic cave walls. The motivation for cave art (to the extent that "art" is the relevant term) remains a mystery, despite intensive study, but perhaps it provided lessons and instructions for cave people.[19]

Before he dove too deeply into the topic of eggs, Newton did write a short "Abstract of Mr. J. Wolley's Researches in Iceland Respecting the Gare-Fowl or Great Auk (*Alca impennis*, Linn)," summarizing their 1858 expedition. This was the article he published in *Ibis* in 1861 in which he told the far simpler tale. Wolley had intended to work with the material gathered in Iceland himself; now Newton went through the *Gare-Fowl Books* and reached some general conclusions. He made comments here and there, conscientiously signed *A.N.*, but he made no attempt to edit or publish the bulk of the manuscript.

He was likely well aware that his and Wolley's experience during the Icelandic expedition could throw light on historic events—that the *Gare-Fowl Books* were a unique source on the fate of the great auk, and on the people who knew the most about the last hunting trips on which great auks had been caught. But with Wolley dead, and the massive work required to catalogue and preserve his vast egg collection, the *Gare-Fowl Books* languished. Other than Newton himself, only a handful of library visitors appear to have consulted the manuscript over the past century and a half, according to the staff of Cambridge University Library, which includes the *Gare-Fowl Books* in Newton's archive. Very many people have been interested in the fate of the great auk, but most (with important exceptions) have regarded Newton's overly neat essay of 1861 as sufficient, and saved themselves a trip to the library.

Eiríkur Magnússon, who assisted Wolley and Newton in Iceland, may well have been one of the few to have perused the *Gare-Fowl Books* for himself in the university library. Perhaps, paging through them upon acquisition, Magnússon noticed the peculiar agreement he himself had made and signed with Wolley and Newton to hunt and collect eight or more great auks on the Great Auk Skerry better known as Hvalbakur. He may have frowned as he recalled, in retrospect, that there were no great auks on the skerry when he arrived, after a long and strenuous horseback ride and a dangerous boat trip, and there had not been any for as long as the local people could remember. For him, there had been little point in making this agreement, postponing his examinations at the clerical seminary, and undertaking a perilous journey—except in that the great auk had introduced him to Newton, and secured his future as a librarian at Cambridge University. Magnússon lived close to Newton for more than four decades. In 1866, Newton noticed an advertisement from him in a London newspaper, offering Danish lessons. He invited the Icelander to visit him in Cambridge and assist him with translation.[20] Later, through his friendship with Newton, Magnússon won the post of librarian at the university, where he remained for thirty-eight years, from 1871 to 1909; he was a prolific translator and publisher of medieval Icelandic writings.[21]

Symington Grieve Steals the Scene

While Newton was hard at work on *Ootheca Wolleyana*, a little-known Scottish amateur ornithologist, Symington Grieve, wrote to him several times asking for information about the great auk. Newton politely replied, sharing his knowledge. On 3 March 1885, Grieve requested information on a stuffed skin of a great auk apparently spotted in Iceland in 1858, adding, "I am not in possession of any of your valuable papers on Alca impennis and if you can oblige me with a set they will be greatly valued."[22]

A few days later, Newton responded, sending a paper he had published in 1870, "On Existing Remains of the Gare-Fowl (*Alca impennis*)." In this brief article, published in *Ibis*, Newton attempted to tally the great auk specimens held by various museums and private collections in several European countries and the United States. He wrote: "Easy as the task may seem, it is in fact very difficult . . . to compile a perfect list of the skins, bones and eggs of this bird which exist in collections." Newton

FIGURE 9.3. Symington Grieve, author of *The Great Auk, or Garefowl*. (Photo: Errol Fuller.)

concluded: "in citing authority for my statements I intentionally limit myself to as much as is sufficient for my present purpose. It is in nearly every case that of an eye-witness."[23]

Later in the year, Newton received a copy of a new book published by Grieve, *The Great Auk, or Garefowl*. It transpired that it was a work of high quality, grounded in extensive work by the author—probably something like the book Newton had envisaged writing himself. Grieve had been corresponding with bird enthusiasts around the world for twenty-five years, asking about the fate of the great auk, dead or alive. He is diplomatic and generous in the introduction to his book, crediting Wolley and Newton for having done "service to science by their labours; . . . the latter has contributed, in several papers, the results of their united work. . . . It is hoped that he may be spared to write a full history of The Great Auk and Its Remains," for which he has the capabilities possessed perhaps by no other ornithologist at the present time.[24]

Newton felt that his work on the great auk had all been for nothing—that another man, hitherto unknown to the ornithological community, had stolen his place. When approached by the editor of Nature to review Grieve's book, he refused at first, suggesting other experts who could undertake the task. None was willing. Still hesitant, Newton even consulted the author himself, who declared that he would be honored if Newton were to write about his book. Their relationship seemed quite friendly, and Newton visited Grieve in Edinburgh. In the end, Newton submitted (or seized the opportunity, as some would later suspect).

The review, published on two pages of the prestigious journal Nature in October 1885, is unusual. Much of the text is made up of citations from Grieve's book, and an explanation of why Newton is compelled to write about it—and why it is hard for him to make a neutral judgment. Then Newton takes aim: He writes that the gravest judgment that must be made of the author is that "he has needlessly raised fresh difficulties for future investigators. Mistakes that have taken years of labor to correct, and the correction of which has been published, are again set agoing, just as if no progress in that direction has been made. . . .

Mr. Grieve has been unable to distinguish between good evidence and bad."[25] The denunciation is harsh, but not backed up by any examples to prove the point.

Grieve's understanding of Newton's review is clearly worded in a letter he wrote when the review was about to be printed: "To some minds the article will suggest a notice such as you sometime see on lands protected for game—'Garefowl Preserve, Trespassers will be prosecuted.'"[26] Newton had laid claim to that territory, attacking the poacher, like a bird fighting for space for feeding or nesting. Despite the fraught relationship between Newton and Grieve, they continued to correspond.

Passports and Censuses

In his review of Grieve's book, as elsewhere, Newton gracefully gave most of the credit for their collective work in Iceland to his companion. Wolley, he wrote, "was living for two months at the miserable fishing-village whence all the later gare-fowling expeditions had started, pertinaciously and laboriously examining, cross-examining, and re-examining every survivor who had taken part in them."[27] Perhaps the reference to the "miserable" hamlet hinted at some tension between guests and hosts at Kirkjuvogur.

While Newton published a couple of informative and perceptive articles on the Icelandic expedition soon after he returned, it seems strange that he didn't dig deeper, with Wolley's manuscript at hand, even leaving aside the intervention of Grieve and the personal blow that his publication of *The Great Auk, or Garefowl* represented. But Wolley and Newton's collaboration eventually did make quite an impact on both the natural sciences and environmental protection, although it took some time for their ideas to sink in.

Newton credited Wolley, for example, for drawing his attention to the novel idea of bird censuses. Prior to censuses, bird-
ringing (also called "banding"), applying tags and numbers to
birds on a large scale, had proved useful for following the movements of individual birds and establishing their geographic
distribution.[28] Some ornithologists thought of the tags as the
"passports" of birds.[29] An international entourage of volunteers
would track and register the birds' journeys across national
boundaries, in a sense granting them multiple residence or citizenship. Now, Wolley reasoned, concerns with declining species
demanded some solid measure of *scarcity*, if not systematic censuses or polling. He first acquainted Newton with his idea in a
letter dated 7 June 1856. In Newton's words: Wolley "stated
that . . . he should like to 'give some account of the British birds,
of which so little is known . . . beyond the bare list.' He wished
to begin 'by naming the birds which are commonest in England
and most characteristic of our bird-fauna'; above all, 'to be able to
represent by numbers the relative abundance of each species.'"[30]

This was a very Victorian project, Newton observed: "Just at
this time, when we are on the eve of taking the human census
of the British Empire, . . . it is not inappropriate to bring a
somewhat similar design as regards our Fauna to the notice of
naturalists."[31] While Wolley left few notes on the subject, he
seems to have been preoccupied by it toward the close of his
life. Prior to Wolley and Newton, American ornithologists had
expressed similar ideas. In 1811, Alexander Wilson (1776–1813),
often seen as one of the founders of American ornithology, con-
ducted surveys of birds in Pennsylvania.[32] Bird surveys were
only launched in the UK in the 1920s, but later they became
routine in many contexts.

Newton's ideas on extinction were equally visionary. As early
as 1778, the Frenchman Georges-Louis Leclerc, Comte de

Buffon, had identified the seventh and final epoch in the history of the earth, "When the power of man has assisted that of nature." Buffon's ideas were not widely accepted (his book was not translated into English until 2018).[33] Evolutionary theory was in its infancy in the late 1700s, and extinction had not yet been established as an epistemic object. Fossils representing extinct species no human had ever seen were just beginning to open a window into the distant history of life on earth. When in Iceland in 1858, Newton must, however, have sensed what Buffon was trying to capture; somehow things were changing for birds and animal life in general, and humans were playing a significant role. Indeed, Wolley and Newton's rationale for their Icelandic expedition—once they gave up hopes of rowing out to Eldey and observing great auks on their nests—was the necessity of depicting and documenting just such a moment. Back at Cambridge, reflecting on the complex engagements recorded in Wolley's *Gare-Fowl Books*, and driven by the progressive ethos of the Victorian age, Newton modeled a space for extinction propelled by humans, what can be called "Newtonian extinction."

Henry M. Cowles has skillfully outlined the development of Newton's ideas, their roots in Victorian science, and their impact in the real world. Newton's work, he argues, reveals "a particular moment in the history of Victorian Britain and in the history of science . . . a moment in which the boundaries between science and sentiment, and between those who did and those who did not have the authority to speak for nature, were being redrawn."[34] The Victorian ethos of progress was triggered by advances in the sciences and by the expanding world of empires. This expansion of Britain's field of vision and experience tended to challenge both stasis and ethnocentrism. Things might change—and they might improve. While fatalism was still common in some scientific circles (some species, it was argued, were doomed to disappear), it was on the defensive. Some extinctions, Newton argued, might and should be averted.

Newton was not acting alone. Scottish geologist Charles Lyell (1797–1875), a friend of Darwin's, broached the subject of extinction, as did Richard Owen (1804–92); writing in 1859, for instance, Owen wondered if extinction was simply a matter of the past or was it something ongoing?[35] Newton's key contribution, however, has often been overlooked. Drawing upon a selective adoption of Darwinian thought, his work signified nothing less than the birth of *unnatural* extinction, driven by humanity, as an epistemic object—a phenomenon to be subjected to scientific inquiry and political action. The combination of the unnatural and the harmful added a new spin to Victorian ideas of extinction. "Lyell's influential interpretation of extinction as consistent with a balance of nature in dynamic equilibrium," American historian David Sepkoski argues, "gave it a kind of positive moral valence: extinction was necessary, and even *good*, for the maintenance of a stable economy of nature."[36] Newton strongly disagreed. From his point of view, catastrophe was in the air—unless humans acted responsibly, and did so in time.

CHAPTER 10

NEWTONIAN EXTINCTION

When Wolley and Newton returned from Iceland in 1858—having never seen a live great auk and having no skins or eggs to show for their grand expedition—they found the natural sciences revolutionized. History had been made in England while they were gone. On July 1, the secretary of the prestigious Linnean Society in London had read aloud eighteen pages, comprising the forthcoming papers of Charles Darwin and Alfred Russel Wallace on the origins of species and the theory of natural selection. Wolley and Newton were in Reykjavík at the time, busily interviewing taxidermists and copying archival material. Only well after their return to England did they learn about this new theory, which placed their inquiry into the disappearance of the great auk into an entirely new context.

In Iceland, stuck ashore, Wolley and Newton had talked for hours about similar ideas. What meaning should be attributed to the concept of *species*? How had different species come into being? "Of course, we came to no conclusion worth anything," Newton confessed.[1] Ideas about speciation and evolution had been floating around in the scholarly world since the eighteenth century, without ever becoming a subject of serious debate or being formalized into a theory that presumably could be proven or disproven.[2]

Upon hearing about the new theories propounded by Darwin and Wallace, Newton was immediately captivated. Here was the secret of species, their birth and instability. He wrote:

> Not many days after my return home there reached me the part of the *Journal of the Linnean Society* which bears on its cover the date 28th August, 1858, and contains the papers by Mr. Darwin and Mr. Wallace. . . . I know that I sat up late that night to read it; and never shall I forget the impression it made upon me. Herein was contained a perfectly simple solution of all the difficulties which had been troubling me for months past. . . . I never doubted for one moment, then nor since, that we had one of the grandest discoveries of the age—a discovery all the more grand because it was so simple.[3]

Three decades later, Newton tried to trace his quick conversion to Darwinism by asking his friend, zoologist H. B. Tristram, to refresh his memory:

> Towards the end of the month (August 1858) appeared that part of the Linnean Journal, and behold all to me became as clear as possible! Such a revelation never was before nor will be, I think, again to me. . . . [P]lease let me have any letters you can find showing my frame of mind at the time, or at least before the publication of the "Origin," which was not until Nov. (or perhaps Dec.), 1859.[4]

Newton told Wolley about Darwin's theory, but Wolley was not convinced, initially at least. He wrote in a letter to Newton that he could not as yet claim to understand the basis of the theory. The term "natural selection," which Newton has used, led him to doubt whether the new theory could be reconciled with his own views.[5]

Darwin and the Fate of the Great Auk

Newton sought out Darwin soon after he returned from his expedition to Iceland, and the two began a lively correspondence. Newton is likely to have mentioned his interest in the great auk to his newfound mentor: both men were interested in birds. Darwin studied pigeons, which he bred experimentally in his garden, and he had observed variants of finch in the Galapagos.[6] In one of his letters to Darwin, Newton states that he had observed for the red-necked phalarope (*Phalaropus hyperboreus*) that the two sexes differed with respect to plumage and devotion to the young and that it had occurred to him, possibly while he was in Iceland, that somehow this was connected. Now it all made sense in the light of natural selection: "Your beautiful theory had not been published & was of course unknown to me, or I should have probably at once seen the desirability of making precise observations to determine the truth of my hypothesis."[7]

But if Darwin and Newton ever discussed the great auk, we have no record of it. The archivist in Cambridge University Library, where Darwin's papers are kept, can see no indication that Darwin ever mentioned the bird—"but you never know; more is always being discovered," she says. She retrieves dozens of letters written by Darwin from flea markets and auctions in England each year. Charles Darwin has not said his last word.

Strangely enough, there appears to be only one reference to an auk-like bird in Darwin's entire correspondence. Richard Owen drew attention to *Alca torda*, the razorbill (now assumed to be the closest living relative of the great auk), in an encyclopedia entry in 1859, when his fierce opposition to Darwin's theory was brewing. The entry concludes with this paragraph, with a hint of irony: "As to the successions, or coming to be, of new species, one might speculate on the possibility of a variety of auk being occasionally hatched with a somewhat longer winglet and a dwarfed stature,—of such an origin, for example, of *Alca torda*;—but to what purpose?"[8]

Newton was probably closer friends with Alfred Russel Wallace, the co-author of the theory of natural selection, and Wallace, if anything, was even more interested than Darwin in birds. Newton and Wallace collaborated within the British Association for the Advancement of Science (BAAS) on the classification and naming of species. Both could envision a period of radical environmental change, like the current Anthropocene, as a result of human activities. They approached the concept, however, from quite different perspectives. While Newton focused on contemporary change and the near future, in particular, on the pending extinction of the great auk, Wallace was interested in the far distant past: on the period when anatomically modern humans migrated out of Africa some 100,000 years ago.[9] Newton's Anthropocene—unlike that of Wallace but, significantly, presaging modern versions—allowed for humanity's awareness of both its impact on the planet and its sense of obligation to redress the balance.

Correspondence between Wallace and Newton was not restricted to zoology. Once Wallace confided to Newton: "I was to have been married in December,—everything appeared serene,—invitations were sent out, wedding dresses ordered & all the programme settled, when almost at the last moment without the slightest warning the whole affair was broken off. I am now getting over it a little, but am not very bright, & cannot tell when I shall go at birds again."[10] They must have talked about birds, then, when uninterrupted by their private lives. Yet, none of Wallace's letters to Newton seem to be focused on the great auk.

Origins, Not Extinctions

Whether they discussed the great auk or not, however, Darwin and Wallace's work placed the world of birds into a new context for Newton, sparking new questions about their future and fate. In 1869, for example, ten years after the publication of *On the Origin of Species*, Newton hypothesized in the first issue of the periodical *Nature* that the colors of birds' eggs were the result of natural selection.[11] Darwin appreciated Newton's support. Shortly after Newton had visited him at his home, Down House, in February 1870, Darwin sent him a letter:

> I have just read what you have said in the *Record* about my Pigeon chapters, and it has gratified me beyond measure . . . for you are the first man capable of forming a judgement . . . who seems to have thought anything of this part of my work. . . . I thoroughly enjoyed the Sunday visit which you and the others spent here.[12]

Although he was in frequent communication with Newton—who sometimes seems to have thought of nothing *but* the great auk—Darwin apparently had no interest in discussing its extinction, whether as a biological subject or a theoretical concept. Gillian Beer points out that Darwin is "scornful about people's surprise in the face of extinction and scoffs at their melodramatic explanations for the event, which he suggests, has usually been quietly coming on through diminution of numbers and loss of habitat."[13] Given such a position, perhaps the case of the great auk was a delicate matter best avoided when Darwin and Newton met or corresponded.

There were other delicate matters as well. When Newton applied for the new post of professor of zoology and comparative anatomy at Cambridge University in 1866, he contacted Darwin to seek his support. Darwin responded that Newton was not the right man for the job, given his training.[14] Newton wrote Darwin a polite reply (though he must have been shattered), and they remained in contact. Newton received support from other directions and was nonetheless appointed to the professorship, by a majority vote of 110 to 82, and became a full member of the Magdalene College faculty (see plate 16).

Though disappointed that Darwin had not helped him obtain the professorship, Newton remained a staunch evolutionist. From the time *On the Origin of Species* was published on 24 November 1859—a date now commonly known as "Evolution Day"—Newton had loyally backed Darwin and Wallace in correspondence with his friends and colleagues, some of whom were resistant to the idea of natural selection. They and other evolution skeptics saw the theory as being in conflict with Holy Writ, in claiming that the human race was descended from apes—that it had not been created fully formed by God in the Garden of Eden, but had developed step by step. Was it possible that the Divine Maker had created imperfect, second-class creatures, destined to become extinct?

Given Newton's own strong religious leanings, it is somewhat puzzling that he became one of the early advocates of Darwin's and Wallace's theory. One reason he found evolution appealing was that it made sense of observed similarities—common ancestry helped to explain archaeological remains and morphological variation.[15] Bone collections were no longer pure wonders of nature; now they received new meaning as windows into zoological connections and the history of life on earth.

Newton was present, though he did not take part, when the controversies surrounding evolutionary theory were hashed out during several meetings in England. That of the British Association in Oxford on 30 June 1860 turned out to be one of the most important public debates in nineteenth-century biology. Newton described the bitter debate between Darwin's supporter, the anthropologist and biologist Thomas Henry Huxley (1825–95), and Samuel Wilberforce (1805–73), the Bishop of Oxford, in a letter to his brother Edward:

> Huxley was called upon . . . to state his views, and this brought up the Bp. of Oxford. . . . Referring to what Huxley had said two days before, about after all its not signifying to him whether he was descended from a Gorilla or not, the Bp. chafed him and asked whether he had a preference for the descent being on the father's side or the mother's side? This gave Huxley the opportunity of saying that he would sooner claim kindred with an Ape than with a man like the Bp. who made so ill an use of his wonderful speaking powers to try and burke, by a display of authority, a free discussion on what was, or was not, a matter of truth, and reminded him that on questions of physical science "authority" had always been bowled out by investigation, as witness astronomy and geology. . . . The feeling of the audience was very much against the Bp. . . . Nothing very particular occurred during the last few days, and I did a good deal of lionising.[16]

When Darwin published his second major volume, *The Descent of Man, and Selection in Relation to Sex*, in 1871, however, the religious Newton showed little interest in the book, compared to *Origin*. In *The Descent of Man*, Darwin attempted to address the complications of including humans in his controversial framework of natural selection.[17] Significantly, Newton was able to maintain his faith in Darwin only by refusing to speculate on the troublesome social implications of his theories.[18] *Descent* turned out to be no less controversial than *Origin*; a hundred and fifty years on, it is charged with offering a racist and sexist view of humanity.[19]

Newton was also present at the meeting at Cambridge in 1862 that signaled the defeat of anti-Darwinism. Here too there was "a grand kick-up," this time between Richard Owen and Thomas Henry Huxley, "the former struggling against the *facts* with the devotion worthy of a better cause," Newton remarked.[20] Owen was one of the most influential biologists in Britain in Darwin's time. Interestingly, while he challenged Darwin's theory at first, a few years later he claimed he had himself conceived of the central idea earlier. Realizing that Darwin had landed the big fish, Owen wanted his share of the glory.[21]

"Probably Mistrusted"

Darwin's and Wallace's theory was a huge step forward in our understanding of the living world. It was now clear that species changed, and that new variants were constantly emerging. The mechanisms of inheritance, however, remained mysterious. In the first chapter of *On the Origin of Species*, Darwin wrote that "the laws governing inheritance are quite unknown; no one can say why the same peculiarity in different individuals of the same species, and in individuals of different species, is sometimes inherited and sometimes not so; why the child often reverts in certain characters to its grandfather or grandmother or other much more remote ancestors."[22]

These mysterious laws of heredity were partly explained in an essay by Austrian monk Gregor Mendel (1822–84) in 1866. Mendel's experiments growing pea plants in a monastery garden marked another historic turning-point—one even fewer people were aware of at the time. The generations of peas systematically cultivated by Mendel would reveal some of the laws of genetics that, along with Darwin's and Wallace's theories, helped to explain the diversity of the living world and gave rise to a scientific revolution in the twentieth century. Before long, heredity was established as a thriving avenue into biological history and life itself.[23]

Darwin apparently remained unaware of Mendel's work throughout his life. His copy of Mendel's paper survives, but it was clearly never read: The pages remain uncut.[24] Not until the 1930s would Darwin's and Mendel's theories be integrated, in the so-called Modern Synthesis, sparking wholesale and controversial rethinking in the life sciences.[25] While Newton did live to see the rediscovery of Mendel's work, he was not impressed. A few years before his death in 1907, Newton nominated a deputy to present his formal lectures; he chose one of his former students, William Bateson (1861–1926), who coined the word *genetics*. In 1900, Bateson had rediscovered, together with two others, the pioneering research of Mendel, now known as "the father of modern genetics." Bateson himself would be named "the brilliant prophet of Mendelianism," by A.F.R. Wollaston in his biography of Alfred Newton; it was, Wollaston adds, "a subject the Professor [Newton] was uninterested in and probably mistrusted."[26]

Irony of Fate

Despite his support of evolutionary theory, it was in opposition to Darwin's views that Newton developed his greatest achievement—the concept of extinction that paved the way for robust animal protection measures and, more broadly, modern environmental concerns.[27] Newton's secret lay in establishing a clear distinction between unavoidable natural extinction—exposed by Anning and Cuvier and theorized by Darwin and

Wallace—and avertible extinction *due to human agency,* such as that faced by the great auk in the mid-nineteenth century.

Newton rejected Darwin's fatalism, which assumed that events would simply follow a "natural" course; at the same time, Newton drove home the point that political action must be taken to forestall the extinction of species wherever possible. Turning the great auks' nesting grounds on Funk Island in Newfoundland and Iceland's Eldey into slaughterhouses, he argued, was unnatural—the killing of these birds was attributable to human beings; the volcanic eruption under the Great Auk Skerry in 1830, which deprived the great auk of its breeding grounds there, by contrast, was subject to natural laws. This dualism—juxtaposing the natural and the artificial—was where Newton and Darwin differed, although they were in agreement on so much else.

Newton's thinking was spurred on, as well, by Darwin's opponent Richard Owen. In *Palæontology,* his summary of extinct species published in 1860, Owen wrote that "ancient" extinction was not just a thing of the past: "admitting extinction as a natural law, which has operated from the beginning of life under specific forms of plants and animals, it might be expected that some evidence of it should occur in our own time." "Reference has been made to several instances of the extirpation of species," he continued, "*certainly, probably or possibly,* due to the direct agency of man."[28] Owen was aware of reports about the disappearance of Great Auk Skerry in 1830 and reports on the killing of the two great auks on Eldey in 1844, but he felt confident to claim that although the great auk seemed "to be rapidly verging on extinction," it had "not been especially hunted down"; the "disappearance of the fit and favourable breeding-places of the *Alca impennis* must form an important lament in its decline towards extinction."[29]

Perhaps it was these statements by Owen that pushed Newton to publish his "Abstract of Mr. J. Wolley's Researches in Iceland Respecting the Gare-Fowl or Great Auk (*Alca impennis*, Linn)" in *Ibis* in 1861. In any case, in this summary of the 1858 expedition, he formally objected to Owen's thesis and set the record straight, stating in a footnote that he was "under the necessity of dissenting from the opinion expressed by Professor Owen."[30] Newton insisted that while the deadly forces of nature might operate equally fast, sometimes even *faster* than artificial ones—the volcanic eruption at Great Auk Skerry being a case in point—they were different from the human forces he documented: "To the destruction which the Great Auk has experienced at the hands of man, must, I am confident, its gradually increasing scarcity be attributed."[31] While Owen trivialized the role of humans, Newton put it at the center:

> I am well aware that nothing but the extraordinary interest that attaches to this bird warrants me in occupying so much space. It must be remembered that it is not merely a matter with which ornithologists only are concerned, but is one of far higher and general importance.[32]

That "far higher and general importance" suggested it was the time to address the damaging impacts of humans on bird habitats.

Newton felt strongly that lessons should be learned from the extinction of the great auk. Those lessons were not solely concerned with the last successful hunting trip to Eldey: even if a few birds might remain alive on some unknown skerry, the species should be seen as "belonging to the past." The extinction of the great auk was a historical fact. But because it had happened so recently, Newton said, "we possess much more information respecting the extermination process, than we do in the case of any other extinct species."[33] Newton's distinction between the natural and the artificial set its mark on all kinds of human activities, both during and after the Victorian age. His words are no less relevant today than when they were penned in the nineteenth century. The human race is part of Nature's creation, Newton said, but that is no reason to sit back and give up in the face of extinction: "It may be said that I have taken too gloomy a view of this matter. . . . I wish I could think so. But I believe that if we go to work in the right way there is yet time to save many otherwise expiring species."[34]

This represented a paradigm change, if not a clear shift. After Newton, the extinction of a species could no longer be thought of as inevitable, positive, long-term, and independent of humans. Newton carved out a space for the scientific management of vulnerable habitats and species, against both unchecked humanitarianism and brutal exploitation, at a time when the Victorian age was rethinking the slippery barrier between elusive sentiment and scientific objectivity.[35] Zoology and the natural sciences had something to offer the world other than discovery, sentimentalism, and cabinets of curiosities.

Foreshadowing modern views, Newton suggested that public policy, informed by natural science, should confront collective extinction, engaging with the existential crisis of mortality at the species level rather than that of individuals. For him, as Henry M. Cowles remarks, "a feeling against extinction was a feeling on behalf of a class of organisms as a whole—usually a species."[36] For Newton, however, the loss of a species was not only to be avoided because of sympathy for the species involved; overall, extinction rendered the job of tracking evolutionary history increasingly difficult.

Newton also deserves credit for the idea that extinction is not a single event. For him, the extinction of the great auk began long before the last creatures died—whether on Eldey in 1844 or not—and the influence of their extinction lasted for a long time afterward. Despite his and Wolley's untiring quest for evidence of the last of the great auks on Eldey island and their painstaking piecing together of the latest successful hunting trips of Vilhjálmur Hákonarson and his crews, Newton understood that extinction was a long process. In 1873, he appeared before a parliamentary committee at Westminster. He was asked—possibly with an ironic, even skeptical undertone, by politicians with limited interest in knowing what extinction might mean—whether he had "observed the habits of birds for a very long time." He steadily replied: "I have paid a good deal of attention to the subject of the extermination of birds in various countries, and the causes that have produced it. Of course, when I speak of the extermination of birds, I also mean the preliminary process; that is to say, making them grow rare."[37]

More than any other scholar, Cowles has restored Newton to his rightful place, brought out of the void and to the fore.[38] It is quite ironic, Cowles points out, that Newton remains more or less unknown today despite his vital contribution to the framing of our modern concept of extinction. The irony of the silencing of Newton is particularly striking in that Darwin, who was certainly not in the vanguard in the battle against extinction but was seen as a leader in the field of natural sciences, had refused to give Newton a testimonial when he applied for the chair of zoology and anatomy at Cambridge. Darwin did not, one may note, totally gloss over extinction caused by humans, although he was primarily concerned with natural extinction. His discussion, however, focused on the disappearance of human "races" (a popular Victorian idea), the declining numbers of indigenous people in the New World in the wake of European colonization—an issue Newton was not interested in.[39]

Updating Alfred Newton

For his concept of extinction, Alfred Newton deserves a place alongside other pioneering environmentalists, such as Alexander von Humboldt (1769–1859). In a series of lectures in Berlin in 1828, and then in a massive treatise in German in four volumes (the fifth one remained unfinished when he died), Humboldt presented the then-novel notion of a web of life connecting everything from subterranean furnaces through living creatures to the planets and the universe. *Cosmos: A Sketch of a Physical Description of the Universe*, based on long journeys and detailed observations, was widely read by naturalists and artists. When the first volume was published in English, soon after it appeared in German in 1845, Darwin professed himself anxious to see it; later Darwin would say that Humboldt's works sparked his travel to distant countries as a naturalist on her Majesty's ship *Beagle* in 1831.[40] Humboldt's ideas also inspired the key environmental thinkers Henry David Thoreau (1817–62) and George Perkins Marsh (1801–82). Thoreau's book *Walden, or Life in the Woods*, was published in 1854. As one of Thoreau's biographers has written, "the railroad whistle across Walden pond sounded the death knell of an old world and the birth of something new . . . when fossil fuels put global economies into hyperdrive, the Anthropocene."[41] Similar ideas were put forward by Marsh, who warned against the damaging impact of humans on other species in his book *Man and Nature*, published in 1862.[42]

About a century and a half later, environmental problems associated with the loss of biodiversity have escalated beyond any linear model Newton or his peers may have entertained. Near the end of the twentieth century, environmental historian Donald Worster, editor of the influential book *The Ends of the Earth* (1988), argued that it was "irresistible to ask whether we are passing from one era into another, from what we have called 'modern history' into something different and altogether un- predictable."[43] That "something different" still had no estab- lished name even in 1988, despite the dramatic "ends" in the book's title. Worster persuasively argued for the need to move beyond histories of countries to a "second history" of planet Earth. He didn't, however, have much to say about extinction, apart from pointing out that "[m]ore commodities for all must mean . . . more crowding, more depletion, and more extinc- tion."[44] The loss of species tended to remain submerged in dis- courses with other labels. A notable exception (mentioned by Worster) was Paul and Ann Ehrlich's book of 1981, *Extinction: The Causes and Consequences of the Disappearance of Species,* which has a few lines about the fate of the great auk.[45]

Now, with the growing awareness of the planet-wide de- struction of habitat, mass extinction deserves and enjoys a se- cure and visible place on public agendas of all kinds. Elizabeth Kolbert's bestseller *The Sixth Extinction* (2014) is a significant landmark. But today, in the Anthropocene, Newton's ideas have their limits—and it's worth exploring why. The nature-culture dualism advanced during the Victorian age established a partic- ular division of scientific labor, separating natural science, the humanities, and the emerging social sciences. While Newton's work helped to place unnatural extinction—"cultural" extinction, in modern terms (arguably it's all natural)—on the political agenda, granting science the necessary authority to speak for rare species and endangered nature, at the turn of the twenty- first century the distinction between nature and culture and the privileged role of expertise advanced by Newton have been in- creasingly questioned. For one thing, scientific observers of nature are no longer seen to be suspended over the realities they observe, but necessarily embedded in them. In the process, the scientific project has, once again, been redefined.

At the same time, the relationship of humans to the planet has been radically altered. A United Nations report leaves no doubt that the rapidly escalating extinction of animal species is a grave and growing global problem.[46] The human race (part of it, more precisely) is making its mark on earth, to the extent that the natural and the artificial have now been more or less conflated. The distinction that Newton advocated, between natural and unnatural extinction, has been increasingly destabilized, as the natural and the cultural have more or less collapsed. Human impact is inscribed throughout the "natural" world. Newton's narrow focus on islands and single species is likewise no longer tenable. In terms of mass extinctions, the whole planet may now be projected as one insular "zoological region," to borrow Newton's term. The earth (or Gaia) "can be seen as an isolated complex of ecosystems surrounded by interplanetary space."[47]

The Unnatural Growth, and Collapse, of the Natural

American anthropologist John Bennett (1916–2005) sensed in the 1970s that an "ecological transition" was brewing: Due to the "growing absorption of the physical environment into the cognitively defined world of human ends and actions," he argued, "there is (or shortly will be) only, and simply, Human Society: people and their wants, and the means of satisfying them."[48] Four decades later, American philosopher and biologist Donna Haraway succinctly described what had happened, outlining the spectacular implications:

> What used to be called nature has erupted into human affairs, and vice versa, in such a way and with such permanence as to change fundamentally means and prospects for going on, including going on at all.[49]

Newton, of course, couldn't foresee such implications of mass extinctions as what anthropologist Deborah Bird Rose (1946–2018) called "double death." In the Anthropocene, she argued, "damaged ecosystems are unable to recuperate their diversity," shattering the "natural" bond between life and death through the flow of generations.[50] As a result, death piles up and spreads like fire, without necessary renewal. Attempts to kill "invasive species," for instance, by means of poisoning, may continue to kill other creatures long afterward. Newton's neat separation of science, politics, and the real world, which served his contemporaries well to frame unnatural extinction and boost the role of the advancing sciences, now needs to be revised, allowing space for more dynamic and democratic approaches encompassing, among other things, collective collapses, chaos, blurred boundaries, and crowdsourcing, always beyond narrow horizons and privileged expertise.[51]

The geologic force of humanity represented by the term "Anthropocene" not only affects the planet "itself" or its crust, it also upsets biological organisms. Nowadays, some people speak of declining "planetary health" as a result of the escalating loss of species. Triggered by the destruction of habitat due to intensive farming and logging and the growth in wild animal trade, the Covid-19 pandemic that began in 2019 had massive impacts throughout the world. The deadly virus may have taken off from the natural habitat where it had been born and contained, but it may also have leaked from a laboratory. One of the human responses to the pandemic, general lockdown and the closing of national boundaries in many contexts, created a "Great Pause," an "Anthropause," which provided an unexpected opportunity to estimate, on a global scale, the effects of human mobility on wildlife, including birds.[52]

Some Anthropocenic impacts leave deep imprints on living bodies, partly due to pollution.[53] Increasingly, seabirds, fish, and marine mammals are threatened by microplastic, which enters their muscles and veins. The results may reduce the organism's chances of adapting, both accelerating the process of extinction and making human attempts to respond ineffective—or, perhaps, simply too late. Such impacts, of course, violate the neat distinction between nature and culture advocated by Newton and his contemporaries. The Victorian question—as posed by Richard Owen, for instance[54]—as to whether the extinction of a species is the result of the creature's own nature (its internal essence or processes) or its habitat, is increasingly redundant. Extinctions are now both natural and cultural at the same time.

In her book *The Human Condition*, German philosopher Hannah Arendt (1906–75) suggested that the advancing biotechnology of the twentieth century—for instance, the attempt to "create life in the test tube"—destabilized the nature/culture divide: "The social realm, where the life process has established its own public domain has *let lose an unnatural growth, so to speak, of the natural.*"[55] Arendt's words foreshadowed not only genetic engineering, the refiguring of genomes and viruses in laboratories, but also the prospect of the termination of life itself as a result of human activities, in the wake of the current environmental crisis. Mass extinction, perhaps, is the example par excellence of an "unnatural growth"—or, more properly, the unnatural *collapse*—of the natural. Grappling with the prospect of the conflation of the natural and the cultural in theoretical discourse and its manifold implications for human understanding and political practice represents one of the major challenges of our age.

To be fair, Newton sometimes used strong words, hinting at what we would now call mass extinction. In *A Dictionary of*

Birds, he defined "extermination" (sometimes leading to extinction) as literally the "driving out of bounds or banishment, . . . a process which, intentionally or not, has been and still is being carried on in regard to many more species of Birds than most people—not excepting professed ornithologists—seem to recognize." He went on, "The inhabitants of islands" are "especially subject to this fate," adding in a footnote that in "some instances the still stronger word, Extirpation, or rooting out, might be appropriately used."[56]

To discuss extinction now demands widening the focus beyond single species like the great auk, to species' relationships and the unfolding web of life. The model organism, if there is such a thing, is no longer a museum item—a stuffed great auk or its egg—labeled and shelved in a neat cabinet of curiosities, but a tree with a vast and open subterranean network of roots normally out of sight. Birds and other beings are radically related.[57] Extinction helps to hammer home the point about relatedness; the loss of a species affects a whole web of life.

Understanding this collective loss, the complex webs involved, and the potential for meaningful action represents one of the major challenges for humanity today. With mass extinction, a diverse range of essential organic connections, formed over thousands or millions of years, will be ruptured or lost.[58] Becoming extinct is necessarily a collective multi-species process; rare species struggle or vanish along with other species and the environment within which they are embedded.

Geoscientists disagree about the definition and exact number of mass extinctions that the earth has already seen. Some have identified no fewer than fourteen (we need not bother with the details), but many refer to the "Big Five" natural catastrophes of deep time. This conveniently carves a space for the impending "Sixth Mass Extinction" of the modern age, a product of human activities.[59] It suggests a division, analogous to that proposed by Newton, between the natural Big Five and the unnatural, human-caused Sixth, during which life gets sucked into vortices of destruction—not instantaneously as in some earlier catastrophes, but nevertheless at an alarming speed in human terms. The mass extinctions of the Anthropocene, and the potential collapse of the web of life, deserve new language and new politics beyond those that Newton imagined or understood. New signature species will replace the great auk as symbols of what we have lost—or perhaps it would be more appropriate to speak, not of signature species, but of signature *habitats*.

CHAPTER 11

THE LAST GREAT AUK

Early in the twentieth century, Mads Peter Nielsen (1844–1931) managed a trading store in the town of Eyrarbakki on Iceland's south coast, one of the most important stores in the country at the time. Like Newton, he had mobility problems; after suffering a stroke, he was disabled for the last two decades of his life. Nielsen was a pioneer in the conservation of birdlife in Iceland; he was a keen egg-collector and taxidermist, and one of the principal benefactors of the Icelandic Natural History Society, founded in 1889. He built an Egg House for his expanding egg and bird collection, a tiny wooden house filled with elegant exhibits; rebuilt in 2004, this hidden treasure continues to attract visitors. In addition to his ornithological work, he studied weather and ocean waves along the south coast. Born in Denmark, he had a keen Victorian spirit.

In 1929, Nielsen wrote an article in an Icelandic newspaper that gave rise to controversy.[1] He commences his piece by recalling that he had once fallen into conversation with a man accused of killing the last great auk. He had started by chatting about commerce and business, then asked the man straight out whether it was true that he had caused the great auk to go extinct. It is not clear where Nielsen had heard this rumor, but the

FIGURE 11.1. Mads Peter Nielsen with parts of his collection in the primary school at Eyrarbakki, Iceland, 1924. Some of the drawings on his wall show great auks. (Photo: Haraldur Lárusson Blöndal. Courtesy of the National Museum of Iceland.)

man's apparent agitated response took him by surprise. Nielsen pointed out that he had not intended any rebuke to the man for his "unfortunate" deed; no doubt he would have left the bird alone had he been aware of its scarcity. Nielsen does not mention the name of the man, but goes on to say, "Sadly, the situation now is, with regard to the great auk, that no person will ever again see a live great auk."

A couple of months after Nielsen's article was published, a rebuttal appeared in another newspaper, written by Ólafur Ketilsson, son of the Ketill Ketilsson who had joined his foster brother Vilhjálmur Hákonarson on his "last successful trip" to hunt great auks on Eldey. Ólafur was a farmer, fishing-boat owner, and leader of the district council. He was said to be an intelligent man; though he had never attended school, he expressed himself well in writing and spoke some German, English, and Danish. Ólafur writes that on reading Nielsen's article, he sensed that it expressed "a bitter resentment and accusation . . . of this man whom he believes was the killer of the last two great auks . . . these accusations . . . are directed at my father's grave . . . for he is widely reputed to have caught and killed the last great auk that were caught" in Iceland.[2] Ólafur maintains that his father was one of the crew, but that the deed itself had been done by another crew member. In either case, he says, it is hardly appropriate to speak of a crime or offense.

Ólafur refers to Wolley and Newton, stating that they had twice traveled out to Eldey with Hákonarson (which is obviously incorrect), but had been unable to land there due to rough seas. "Yet these Englishmen stayed right through until August. They crawled, like worms in the earth, through all the old refuse heaps they found here, in search of great-auk bones, but that search is said to have yielded little result."[3] Given the harsh tone of Ólafur's comment, it is tempting to conclude that his father was the "one exception" to which Newton referred in his summary article about the Iceland expedition, the unenthusiastic crew who stood out, perhaps refusing to reflect and to elaborate on his difficult hunting experience. Ólafur may have sensed his father's ambivalence or irritation over the years when someone mentioned the visit of the British naturalists. What, if anything, made Ketilsson's position different from that of the rest of those interviewed?

Nielsen responded to Ólafur's article, reminding him that he had made no reference to the name of the man he had spoken to years before, nor where he was from. His son, he says, has now revealed these details, naming his father unnecessarily and perhaps contrary to the wishes of everyone involved: "Such a revelation surely comes from where it was least expected, and I am of the view that it is perhaps overstated . . . for that which was widely rumored seventy or eighty years ago and has never been recorded in writing has no doubt been forgotten, and was better forgotten . . . and there is no reason to bring it up now, least of all if it should stain the memory of a good man."[4] Nielsen adds that, since his article was published, he has been asked by many people who the man was; the hunter's name can surely never have been widely known.

Nielsen notes that he has "unfortunately not read Professor Newton's essay, but read many citations from it." An article in a Danish newspaper, he writes, quotes Newton as saying that Brandsson and Ísleifsson had chased down the two birds, but adds that "the man that handled the last great-auk egg—which may well have been so close to hatching that there was life beneath the shell—is said by Professor Newton to have been Ketilsson by name. Hence it is far from unlikely that it was indeed Ketilsson who killed the last great auk!"

Who Killed the Last Great Auk?

This narrative has taken on a life of its own. In *Who Killed the Great Auk?* (2000), Jeremy Gaskell concludes that while in Ketilsson's own account of the event, as recorded by Wolley, he picked up the egg and put it down, it "is more likely that [Ketilsson] dropped it but was too embarrassed to say so."[5] One notable website (under the banner of the John James Audubon Center at Mill Grove) goes further, referring to "Ketilsson smashing the egg with his boot."[6]

There is nothing in the *Gare-Fowl Books* to support such claims. In addition, Newton's role is clearly misrepresented in

Nielsen's reference to the Danish newspaper article: "Professor Newton had been on the hunting expedition in 1844, when the last two great auks were killed."[7] Newton would surely have turned in his grave had he known that he was accused of being personally involved in what may have been the final execution. Like Newton and many other collectors of eggs and skins, Nielsen was hypocritical. While he pioneered bird protection in his coastal community—and in Iceland, more broadly—and he was passionate about the fate of the great auk and rare species, he was an energetic egg-collector himself, bidding generous rewards for the eggs of species in serious decline, even those of the endangered falcon. The price he offered for a falcon's egg was twenty times that of a raven.[8] That must have been difficult to resist.

The dispute in 1929 between Nielsen and Ólafur Ketilsson concluded with a newspaper article by the latter.[9] The controversy no longer focused on the killer of the last great auk. Ólafur now shifted his attention to the nationality and dignity of persons and birds, the struggle of poor farmers and fishers for survival, the Icelandic campaign for self-determination, and the dream of progress and authority over all that belonged to Iceland. Did it matter that Nielsen, former manager of a trading store, had been born in Denmark and had a foreign family name? Ólafur sees nothing to regret in the loss of the great auk. He is hardly likely, he says, to agree with Nielsen "about how 'unfortunate' it may be deemed that the great auk has now, in most people's view, disappeared for good from this world—a bird that was never seen on the coast of this country, but always kept to distant islands and skerries." Ólafur concludes by remarking that trips to hunt the great auk had cost many seafarers' lives in past centuries, and that the bird did not deserve the attention and concern so many had shown it. In a sense, Ólafur refused to grant the bird citizenship, an Icelandic passport.

Ólafur's attitude is interesting because his father was the only witness "examined" in the *Gare-Fowl Books* who displayed any emotional reaction to the hunting of the birds. On Eldey, chasing the birds, "Ketill's head failed him and he stopped," as Wolley and his interpreter put it, as if he could not bring himself to kill a great auk. Perhaps Ólafur's bitter remarks arose precisely from the fact that the last hunting trip had traumatized his father. It is not unlikely that many other people shared his view. The great auk thus became the subject of a vituperous dispute at a time when Iceland had gained autonomy as an independent state within the Danish Empire, and not long before the foundation of the Republic of Iceland in 1944.

Clearly, Ketill Ketilsson was protective of his good name. He built a wooden church in 1864 for the local congregation and later, when the community had expanded, in 1877, a larger stone church that still stands. One recent commentator reasons (without proof) that the two churches "were, in some ways, built to memorialize the events at Eldey in 1844."[10] Ketilsson was only twenty-one when he took part in the controversial hunting trip; he was then working as a farmhand for Hákonarson at Kirkjuvogur. He went on to become a farmer in his own right at nearby Kotvogur. Ketill was regarded as "the biggest farmer . . . in his time" in the area, "and he had the ways of a great man," according to *Icelandic Biographies*.[11]

Yet for some reason Ketilsson's name came to be identified with the killing of the last great auk—beyond the only evidence available, which is that contained in the *Gare-Fowl Books*.[12] Reykjavík engineer Vífill Oddsson, for instance, is descended from both Ketill and Vilhjálmur Hákonarson. People sometimes remarked to him with a sneer that his ancestor had taken part in the last attack on the great auk. He remembers having pestered his grandmother with questions about the killing of the birds.

FIGURE 11.2. Ketill Ketilsson of Kotvogur, member of the great auk hunting crew of 1844. (Photo: Sigfús Eymundsson. Courtesy of the National Museum of Iceland.)

She impressed upon him and his brothers "never to shoot any living thing unless they could eat it," and absolutely no protected bird species. This rule was always enforced and observed by her.

One time, Oddsson asked his grandmother whether it was true, as he had heard, that his great-grandfather Ketill had been

FIGURE 11.3. Ruins at Kotvogur, Hafnir. (Photo: Gísli Pálsson.)

on the last successful hunting trip.[13] She was a punctilious woman who never spoke ill of anyone, but on that occasion she snapped at him. She said of her father-in-law, "He was just there as a youngster, brought in to assist the hunters. He did not kill the last bird." Her account, says Oddsson, is confirmation that "Ketill had not the heart to kill birds on the rock under Eldey . . . that was the way it went." Indeed, "his head failed him."

But the issue of Ketilsson's role and reputation refuses to go away. In 2020, a newspaper article protested persistent clichés about Ketilsson being the ruthless killer of the last great auk. The author was Ólafur Bjarni Andrésson, great-grandson of Ketill Ketilsson; significantly, his article bears the title, "I Still Want to Be Ketill."[14]

Lost at Sea?

While it was widely assumed after Alfred Newton published his 1861 "Abstract" of his and John Wolley's expedition that the last of the great auks had indeed been killed in Iceland in June 1844, some sailors, collectors, and ornithologists were skeptical, both in Iceland and elsewhere. Perhaps there might be a few still wandering the North Atlantic. Newton sometimes received communications that kept his hopes alive. In 1872, for instance, he was sent a card by an Icelandic mariner, who wrote:

> In the year 1869 on 27 June I was aboard the ship *Felicite* near the island of Skrúður off Eskifjörður [East Iceland], where I saw two great auk, which due to storm and rough seas I could not capture.
>
> In Lerwick, 14 May 1872
> S. Thorsteinsson
> Sea captain from Ísafjörður, Iceland.[15]

Even if, in fact, it was completely extinct, the bird remained present in people's minds for a long time. In Iceland, people were on the lookout for great auks until as late as 1894. In that year, Icelandic Captain Hjalti Jónsson heroically scaled Eldey island—the first man to reach its summit—with two companions. The men were partly in search of great auks; they climbed up where the last great auks had been seen. Perhaps there might still be some on the island.[16] Jónsson and his companions were rowed out to Eldey in a six-oared boat. They started by assessing conditions: "On the island there is a small point or sloping rock that projects out into the sea to the northeast. When there is no surf at the island, sure-footed men can easily get ashore there. It was there that the great auk, which was flightless, had its territory on the island—and precisely above that point appeared to be the most promising route up."[17] They looked around for great auk "on the very lowest ledges," but saw none: "No such luck."

Above the climbers was a near-vertical 260-foot cliff. They inched their way up, and started to catch birds, mainly gannets

(*Morus bassanus*) and little auks (*Alle alle*), and to collect eggs: "Hjalti jumped into the horde of little auks. They were unable to get air beneath their wings to take flight, and Hjalti wrung the necks of three hundred and twenty birds. And the men took quite a lot of eggs."[18] This was neither subsistence farming nor collecting for museums, respecting some kind of game law or striking a balance between human needs and sustainability; the scale and brutality of the hunt was more in line with industrial trawling, a sign of the times by the end of the nineteenth century. But, for their bird-catching exploits, Hjalti Jónsson and his company were given a heroes' welcome in Reykjavík: speeches were given, followed by "the biggest and most general drunken binge that Hjalti had ever seen. The Governor was informed that the Danish flag had been raised on Eldey, and it had been claimed as part of Iceland."[19]

Many sightings of great auks were reported after 1844 on North Atlantic skerries in Iceland (1846, 1870), Greenland (1859 or 1867), Newfoundland (1852, 1853), and northern Norway (1848).[20] Some of the reports were certainly apocryphal: people had mistaken another species for a great auk, or had seen what they wanted to see. Others were deemed credible and were probably true: evidence of a few dispersed pairs of birds continuing to breed on islands or skerries for a few years. Such tales were often unjustly dismissed, and unnecessarily strict standards of proof and corroboration were applied. The consensus among scholars today seems to be that the last living great auk was seen off Newfoundland in 1852.[21]

One of the great auk tales deemed most credible, after the species was generally regarded as extinct, is about a bird seen at Vardø in Finnmark county, northern Norway, over a hundred and fifty nautical miles north of the Arctic Circle. Could it be true? Newton surely did not believe it. He took the view that, as no great auk bones had been recovered from refuse heaps north of the Arctic Circle, the bird was not an Arctic species, and hence it would be pointless to seek live birds in the far north. This was the repeated theme of ornithologists, who reiterated again and again that the great auk was *not* a penguin—although seafarers often called it one. The term "penguin of the north" and the Latin name *Pinguinus impennis* were misnomers, based on a misunderstanding. The great auk was not related to penguins in any way, they insisted—rightly, if by "penguin" you mean only the southern species and ignore the long history of the name. Reports of great auks north of the Arctic Circle were treated with skepticism, at any rate.

Forensic Insights

Once it seemed clear that the last of the great auks were dead, museums and collectors around the world scrambled to acquire skins, eggs, and bones of the extinct bird. The Victorian obsession with collecting was past its peak, but anything relating to the great auk remained a prize. There are some eighty stuffed great auks in collections around the world, and an unknown number of preserved skins and viscera. Only about twenty-four complete skeletons exist, while thousands of loose bones (some with knife marks) are kept in museum collections. The skeletons do not have the visual appeal of the taxidermied birds, mounted to look so lifelike in their full plumage. But the bones—what Wolley and Newton termed "relics"—tell a long and complex story of their own.[22] Last, about seventy-five great auk eggs are believed to be extant today, with the vast majority documented and numbered.

What happened to the remains of the two birds caught on that "last successful trip" in 1844? Newton's archives include a letter he received in 1873, tracing the first steps of those two bird-skins from Eldey. The letter reads:

> From the Factor Hansen in Iceland the late Prof. Eschricht in 1844 received two skins of Garefowl but the bodies that had been in them, viz a male and a female of which Prof. Prosch, by that time Curator of the Zoological Museum made a series of preparations viz: throat and lungs; belly and gut-canal; heart and eyes; the sexual organs, male and female; besides the kidneys; ... prepared and placed in 10 vessels and at current kept in the ... collection of Prof. Steenstrup.[23]

Daniel Frederik Eschricht (1789–1863) was professor of anatomy at the University of Copenhagen. He had studied in Copenhagen and Paris, where his mentors included Cuvier, who had opened up new perspectives on geological time, with the help of Anning and her fossils.

Surprisingly little is known of the fate of the two great auk skins that passed through Eschricht's hands, even though collectors generally kept detailed records of all matters relating to great auks. But similar inquiries have yielded fruit. In 2013, for example, it proved possible to solve the mystery of two great auks caught off Iceland around 1833, by carefully trawling through the storage facilities of the Bavarian State Collection of Zoology (ZSM) in Munich.[24] A curator inadvertently discovered the auks in a storage box labeled "owls," kept in a nonornithological part of the ZSM. One suggestion is that the bird-skins from 1844 may have been sent from Copenhagen to Hamburg, and then on to some other destination. Where are they now?

Generations of ornithologists have wrestled with that question, delving deep into archives, but no one has reached a definite answer. Eschricht's and Prosch's papers at the museum in Copenhagen contain a wealth of information on the two men, as well as on the museum itself, but no apparent reference to the two great auks from 1844. Geneticist Jessica Thomas and biologist Thomas Gilbert, together with several colleagues, have been seeking a solution in genetics.

It was Errol Fuller who set the researchers on this path: as author of the key reference book *The Great Auk* (1999), he was familiar with the stories of the various great auk skins that were possible candidates. He took the view that five skins—in collections in Brussels, Los Angeles, and the German cities of Bremen, Kiel, and Oldenburg—were most likely to be the missing ones from 1844, based on the available information on their age and provenance. Fuller had previously maintained that the birds in Los Angeles and Brussels were the most likely contenders.[25] A well-known Belgian taxidermist, Pierre-Yves Renkin, had identified the bird in Brussels as the real deal: the last of the great auks.[26] But all five of these birds had been caught at around the same time. A label on the Brussels exemplar is dated 1846. The museums involved had all contributed to the extinction of the great auk by encouraging hunting of the species at a crucial time. Hence, they were more than willing to cooperate. The study might offer the opportunity to fill in gaps in their catalogues, and to correct possible errors. No conscientious museum curator can tolerate confusion or sloppy inaccuracy.

The genetic researchers acquired tissue samples from the organs of the two birds stored in jars of alcohol in Copenhagen, including shreds from the heart muscle of the female bird; now her heart beats once more, if only metaphorically. Samples were also taken from the five taxidermied birds, in the hope that a comparison of the genome might throw light on the lost birdskins. This research group worked with aDNA—that is, ancient DNA—using technology that had been developed over the past three decades in studies of the history of ancient organisms, among other things in exploring the origins of humanity. Now, the two birds caught on Eldey in June 1844, by the crew of Vilhjálmur Hákonarson of Kirkjuvogur, were back in the spotlight, in an unexpected role.

In a sense, the scholars analyzing the DNA from these samples are in a similar position to that of John Wolley, who questioned the crew members from Kirkjuvogur—although the modern researchers are not left hanging around for weeks on shore, far from home, sustained only by a constant flow of coffee from their hosts. The findings of their genetic study were published in the periodical *Genes* in 2017, followed by a doctoral thesis the next year.[27] It should facilitate the search for the missing birds, say the authors, that the genetic diversity of the great auks has been shown to have been considerable right up until the extinction of the species.

The female bird whose organs are in Copenhagen remains a mystery. Unfortunately, one of the two DNA samples taken from the organs of this bird proved to be unusable; the other, from the heart muscle, turned out to be only distantly related to the five bird-skins that were being studied. Hence, the skin of this female bird has not yet been traced. It has been suggested that her skin might be in the Museum of Natural History and Science in Cincinnati, Ohio, but that supposition has not been proven.

The male was a simpler matter. The genetic comparison revealed that the great auk skin in the Museum of Natural Sciences in Brussels and the samples from the male viscera in Copenhagen were probably from the same bird. The other four birds were only distantly related. In some bird registers, the great auk skin in Brussels is catalogued as number 3 or 13; Symington Grieve and others record it as RBINS5355. Professor Eschricht in Copenhagen probably placed the bird in the hands of a certain Israel, who was a well-known great auk dealer with a strong connection to Iceland. After that, the bird appears to have passed from one person to another: to merchant Linz in Hamburg, then to merchant Frank in Amerstad, until it finally came in 1847 to the then-director of the Brussels museum, Bernard Du Bus Ghisignies (1808–74). It was certainly an eventful life-after-death for the bird.

"As Extinct as the Great Auk"

As professor of zoology and anatomy at the University of Cambridge, Alfred Newton continued gathering everything he could lay his hands on about the great auk until the end of his life. He was constantly working—mostly on producing the four fat volumes of *Ootheca Wolleyana* (which includes six great auk eggs)—to the extent that some of his colleagues saw him as an eccentric. In his own opinion, on the contrary, he was never accomplishing enough; as he recorded of one day's productivity, "a few lines written amid an abomination of desolation." At the same time, he became more and more involved in environmental activism, trying to bring the public's awareness to the fate of birds.

Newton's biographer, A.F.R. Wollaston, observes for one of the "seminars" or "conferences" at Magdalene College: "None of those who were present are likely to forget the occasion, one evening in Newton's rooms, when a young man interrupted . . . with the rather large question: 'Why *do* birds become extinct?' The Professor replied without hesitation, 'Because people don't observe the Game Laws.'" "The conversation languished after that," Wollaston continues, "and we soon returned to our various colleges." The disciples would spread the news and the message.

Newton lectured on the necessity of updating Britain's game laws, founded a society for the protection of birds, and helped draft legislation on bird protection which—despite its hypocritical bias toward aristocratic "sportsmen" and collectors like himself—was a milestone in the field. He wrote or co-authored policy reports and supplements, popular articles in the national press, and scientific essays in important journals, mostly on the protection of indigenous animals, notably birds.[28] He also launched the ornithological periodical *Ibis*. Although he had no secretary, he kept in touch with a great number of scholars and environmentalists. His address book contains several hundred names, and "it can be believed," says Wollaston, "that the accumulated correspondence of more than fifty years amounted to tens of thousands of letters."[29]

Until 1882, Newton also continued to travel widely, despite the lame leg that had afflicted him since childhood. In that year, he visited the small island of Heligoland off Germany to attend an ornithological conference. On the journey, he suffered another injury: boarding ship, he tore the ligaments in his good leg just above the knee. From that time, he could get around only on crutches. His daily walk to work at Cambridge University, his biographer remarks, "was an affair which called for no little effort on his part, for his great lameness—much accentuated by the accident . . . —made him a 'four-legged man'; he used a stick in each hand and his rate of progression was not rapid."[30]

Cambridge University is, and always has been, unusual in certain ways. At this ancient institution, faculty members are based in many individual colleges, where students and staff from different disciplines mix socially over lunch or dinner. For the dons, the colleges were like family, as they lived largely "in college," in the same buildings with their students. Magdalene

College was Newton's only home from the time he first arrived in Cambridge as an undergraduate, with the exception of his time spent traveling. He lived for many years in Old Lodge, a grand building. In his rooms, he offered company and conversation, smoked one of his many pipes, and drank port with friends and colleagues—when he was not busy working.

Life at college in the nineteenth century had much in common with the monastic life: Within the walls that separated students and faculty from the growing town outside, strict rules of discipline and religious observance applied. Formal attire was mandatory—faculty members invariably wore academic robes—and a certain tension existed between students and college authorities, who imposed penalties for even minor infractions. The college gates were locked at 10 p.m., and students who stayed out later were fined. Many students appear to have had limited interest in their studies; it proved difficult to maintain organized teaching schedules—and the results were as might be expected.

Newton's students praised him for always treating them as equals. But he did not enjoy teaching, and he had no ambition to excel in that line of work, avoiding the classroom as much as possible. His lectures were said to be extremely dry and formal. He wrote them out by hand, and read them aloud word for word, sitting at his desk insensitive to context. Occasionally, he paused, reached for his glass, and took a sip of water, as Wollaston, his biographer and former student, notes, then "steadily and relentlessly read on."[31]

He was opinionated and prickly. As he pronounced on one occasion, "the world is made up of trifles, and from some the more we can free ourselves the better. Of this kind are trinominals, motor-cars, hymns, and cats—the last perhaps the worst of all, for there is no avoiding them. Until I am run down by a

THE LAST GREAT AUK 219

motor-car I shan't much mind, and when I am run down I suppose I shall be finished and so mind still less."[32] As Newton approached the end of his life, seriously ill and grieving for friends
who had passed away one after the other, he protested against
dying in his bed, insisting that he should be placed in his arm-
chair: "Here will I meet my fate." He died in his chair on 7 June
1907. Wollaston remarks that Newton "must be accounted an
extinct type, as extinct as the Great Auk or the Dodo of which
he loved so much to write. Such strength of individuality."[33]

British ornithologist Tim Birkhead suggests that while Newton was deeply averse to theorizing, "what he did do—and for
which he should be hailed a hero—was to launch the enterprise
of bird protection."[34] "Overall," he and Peter Gallivan conclude,
"it appears that Newton was a powerful early force in the conservation of birds, a role that today seems largely unrecognized."[35]
While his ideas sank in, his name was often forgotten in the
shadow of larger names in the rapidly expanding field of biology, including those of Darwin and Wallace. Recently, however,
this has begun to change. A series of works have amplified Newton's contribution and placed his work on the environmental
agenda. Birkhead, in *Birds and Us* (2022), argues that the Icelandic expedition of 1858 "shaped much of Newton's subsequent
career. Most important was his realization that extinction was
a topic open to scientific investigation, for at that time it was
almost unconceivable that an entire species like the Great Auk,
once so abundant, could be totally eliminated by the hands of
men."[36] Michelle Nijhuis similarly suggests in *Beloved Beasts*
(2022) that the "British ornithologist Alfred Newton, a contemporary of Darwin's, was one of the first Naturalists to recognize that human-caused extinctions—what he called 'the extermination process'—could bring an entire lineage to an abrupt
and permanent end."[37]

De-extinction

Now and then over the years, species have been said to suddenly reappear, after having been thought long exterminated. Several birds have been confirmed to be such so-called Lazarus species, including the Bermuda petrel (*Pterodroma cahow*), which scared Spanish explorers away with their eerie calls. Considered extinct for three centuries, it was rediscovered on one of the Bermuda Islands in 1951. Also, the flightless takahē (*Porphyrio hochstetter*) of New Zealand, which was claimed extinct late in the nineteenth century, reappeared in 1948.[38] In recent years, with intensive searching, social media, and growing awareness of the threat of mass extinction, such reports have escalated. The great auk as a Lazarus species, however, can be ruled out.

Charles Darwin made the point that species swept away by history would not return. They were gone for good. In *On the Origin of Species*, he wrote: "We can clearly understand why a species when once lost should never reappear, even if the very same conditions of life, organic and inorganic, should recur."[39] This has long seemed blindingly obvious. No doubt many people have wondered why Darwin saw reason to state it at all. But perhaps his words were necessary at the time. Ideas about extinction were unfixed, and Darwin may well have felt it was time to dispel the fantasy of the resurrection of species.

Alfred Newton, on the contrary, entertained the idea that extinction processes could be reversed. And in our own time, discussions of the renaissance, even resurrection, of species is taken for granted—as if Bible stories and the natural sciences had coalesced into one, after centuries of enmity and conflict.[40] Will we live to see the resurrection of *Pinguinus impennis*? Might genetics do the trick?

In the spring of 2015, a group of like-minded people met at the International Centre for Life in Newcastle, England, to discuss the possible reanimation of the great auk.[41] The meeting was attended by more than twenty people—including scientists and others interested in bird conservation—and addressed the principal stages of "de-extinction," from the sequencing of the full genome of the extinct animal to the successful releasing of a proxy animal population into the wild. They were interested in resurrecting the great auk quite literally, to see it thrive once more, in zoos or even on the skerries and islands of the North Atlantic.

Thomas Gilbert, a geneticist at the University of Copenhagen, was one of the scientists who attended the meeting. Not only was he a member of the team who "found" the last male great auk in Brussels, with the aid of genetics, he has managed to sequence the great auk genome.[42] The de-extinction of a species, however, has proved to be a more complicated issue than was originally anticipated—both technically and ethically.[43] Gilbert points out that a re-created species can never be exactly like the original, and that the question must be asked: What counts as "near enough"—ninety-five percent, ninety, . . . ? If the element that is lacking, though it may only account for a few percent of the genome, turns out to be crucial, and makes it harder for a re-created species to survive, nothing has been gained. A re-created great auk that could not swim, for instance, would not be "near enough." Likewise, a great auk capable of flight might be "way too much." For most people, whatever species concept they subscribe to—and there remains a thriving philosophical debate on that subject—a flying bird would hardly qualify for legitimate membership in the species of the "great auk."[44]

Yet a substitute bird that could swim would be welcomed by many, as it might fill in the large gap left by the great auk's extinction.[45] A substitute species might contribute to the rewilding of the oceans, a task that has barely begun; indeed "the underwater realm has been trailing behind its terrestrial counterparts."[46] Interestingly, this idea echoes Philip Henry Gosse's historic aquaria project, reversing the arrows, from land to sea, and operating on a much larger scale. The grand aquarium of the planet's oceans, including the recently discovered seabirds' hotspot in the middle of the North Atlantic, the idea goes, could be repopulated by relatively large charismatic animals, territorially raised and later released into the oceans, where they would be managed and monitored by human divers. Gosse would be amused.

The expense of such de-extinction is high, however, and it is hard to decide which species should have priority: the mammoth? the dodo? the great auk? or perhaps one of the numerous species of tiny snails that rarely generate human concern? It's tempting, and productive, to focus on tall birds and charismatic megafauna, but invertebrates such as snails and insects, which make up most of the animal kingdom (perhaps 99 percent), deserve attention too.[47] In the Anthropocene, this age of mass human-caused extinctions, the selection of species is clearly an urgent, but difficult, concern. The re-creation of the great auk assuredly has symbolic significance, not least in light of the attention the species has garnered from both scholars and the public since its demise. The excessive price nowadays of great auk remains is significant too. In January 2023, a great auk egg sold for $125,000 at Sotheby's. But bringing the bird back to life is a gigantic challenge, if not an impossible one. Perhaps the funds that would be spent on the de-extinction of the great auk might be better spent elsewhere.

AFTERWORD

ON THE BRINK

In 2021, the people of southwest Iceland, the site of John Wolley and Alfred Newton's 1858 expedition, were unexpectedly reminded of the planetary forces capable of generating extinction. The entire Reykjanes peninsula, which had remained calm for well over a century, was shaken by tens of thousands of earthquakes, some of them quite powerful. Sometimes the ground gently danced for several minutes, making people in the towns and countryside facing Eldey terrified. Some locals escaped to inland communities, saying they were seasick at home. Within weeks, an eruption started along the same geologic fault line as the one that sank the traditional great auk breeding grounds on Geirfuglasker, the Great Auk Skerry, in 1830. Although the 2021 eruption ended after a few months, the following year a new crater opened up close by.

These were spectacular eruptions, attracting crowds of tourists. One day, I walked along with them for several hours on a strenuous trail, to see the erupting volcano up close. Watching the glowing lava in the twilight, in the company of a silent group amazed by the view, I pondered again the fate of the great auks— and the resulting learned debates in the nineteenth century about the relative importance of nature versus culture in driving the species to extinction. For several months, as I was finishing the writing of this book, it felt like I was living in two parallel yet strangely related universes; the drama of the last great auks was matched by the threat of an active volcano so close to Iceland's capital city.

Some months earlier, as I was writing about cabinets of curiosities and obsessive Victorian egg-collectors, I suddenly felt pangs of grief for my childhood *Wunderkammer*—probably one of the most modest egg collections of all time—the cardboard box in which hay cradled my long-lost eggs. What is to be done? I locate an Icelandic couple who make replicas of eggs of various species. I give them the dimensions of the average great auk egg (about 4.9 by 3.0 inches) and ask them to make one for me. One day, I receive notification of a parcel awaiting me at the post office. The box turns out to contain a great auk egg, carefully packed to protect it from damage. It is made of plaster, and beautifully colored, copying the markings of an example in some egg collection somewhere. My fingers will not meet around the egg. While only a replica, it is a delight to see—a memorial to a species that was hunted to extinction.

Species Talk

What is lost when the last exemplar of a species dies? What is missed—and what is possibly recovered when a creature is rescued from extinction? It is not only the organism in the strictest sense; while most of the creatures most of us are familiar with have an enveloping layer of skin, it is not always easy to say where the creature ends and its environment begins. The environment is inside the creature's skin, too—in the case of communities of microbes that occupy digestive systems—and, obviously, the creature dwells in an environment.

One of the flaws of Darwinian theory, as American evolutionary biologist Richard Lewontin (1929–2021) forcefully argued, is the assumption of "autonomous . . . environments independent of living creatures."[1] Organisms do not just adapt to their environments, they co-evolve with them. Etymologically, the word *environment* signifies "that which surrounds," but the surroundings and the beings surrounded are always somewhat unstable, always on the move. As Vinciane Despret points out, "we should assume that the milieu or environment itself . . . 'behaves,' that it *allows itself,* or refuses, *to be appropriated. Space co-opts modes of attention.*"[2] In many contexts, as our climate changes, spring and summer now offer quite different modes of attention to those they offered in the past. Importantly, with escalating species extinction, co-opting no longer assumes a seasonal migratory swing; rather, it is permanently reshuffled in the process of adaptation or, more appropriately, the failure to adapt quickly enough. In the Anthropocene, we can anticipate a world in which many species are continuously spiraling out of existence, along with their habitats.

Another species issue is the concept itself. Darwin, it has often been pointed out, failed to develop a clear definition of species, despite the central relevance of the term in his *Origin*; Wallace, in contrast, offered some clarifications.[3] Many later scholars have simply avoided or dismissed the issue. At the outset of their book *Extinction*, American biologists Anne and Paul Ehrlich state that it's best to ignore the complications and arguments in the literature: "there is no need to define species more precisely than as distinctly different kinds of organisms."[4] Dozens of definitions of a species, however, have been proposed, and species concepts remain central issues in biology, anthropology, philosophy, and several other disciplines. Arguably, it is vital for any attempts to preserve biodiversity to agree on what counts as a species: for instance, how many kinds of birds are there?[5] Modern classifications of species, we need to keep in mind, are changing, partly with the application of molecular genetics.[6]

Some scholars see species as one of the Big Four concepts describing the world of life, along with gene, cell, and organism.[7] Others argue that the species concept is so vague that it should be abandoned for a post-species age of fully rankless life-forms, suggesting that there is no need "pretending that everything important happens at a single depth (i.e., the level formerly known as species)."[8] Still others argue that the species concept continues to play a vital role in biological discussions, pointing out that its very ambiguity and fuzziness, like that of the gene, makes it most useful.[9] The variety of species concepts has not obstructed scientific progress. Nor does it strain environmental practice; concerns with diversity are not necessarily species-specific, although obviously they do pay attention to differences between life forms.

Perhaps the strongest argument for maintaining some kind of species concept is that a folk notion of kinds is quite common, if not universal, and is applied to many practices and projects.[10] Throughout the ages, humans have found it important to broadly capture the biological diversity of their habitat, partly for subsistence and mythological purposes, by subsuming family resemblances of some kind—despite great variety, even monstrosities—under a single type, with or without naming or rigid classification. Broadly naming the "kinds" allows not only for authority and control but also for loss and grievability.[11] While the Linnaean system of classification became firmly established in scientific realms, people throughout the world had invented their own systems. Anthropologists of ethnobiological bent have been busy documenting and analyzing them for a century or so. For a long time, these systems have been downgraded as second-class, at best as cultural curiosities shedding light on thinking and context. But, of course, the Linnaean system too was embedded in cultural context, serving specific purposes. After all, there turns out to be considerable agreement between the Linnean fivefold scheme (from the general "living things" to specific "kinds") and (other) "folk" taxonomies. Both assume about five hundred species (Linnaeus had slightly fewer).[12] Sometimes, however, there is disagreement about classification and descriptive details. The members of what modern science and taxonomy call a "species" are not always and everywhere considered to be the "same." The complex geography and history of labels applied to birds now called "great auks" is a case in point.

Sometimes, images of natural kinds may appear as deliberate conceptual collage. Some human (and probably Neanderthal) figures in Paleolithic cave art have animal heads, testifying to both a mythic and classificatory role of the imagery, perhaps indicating metamorphoses along the passage from the underworld to the outside and from the past to the present. Such "collage" may apply to representations of seabirds at Cosquer in Southern France. Anne and Michael Eastham argue that some of the animal kinds are "indeterminate, *with attributes of more than one species conflated into one image*"; as a result, they suggest, "it is not appropriate to be precise in identifying Paleolithic images. An identification of the . . . Cosquer panels as auks or as seabirds nesting colonially is more reasonable than assigning them more narrowly to *Pinguinus impennis*."[13] A similar ambiguity in identification is evident in studies of auk-like images in other Paleolithic caves, in Spain and Italy.[14] One of the birds, however, in the Romanelli Cave, Italy, has three short lines close to the eye, an almost unmistakable emblem of the great auk.

Are species, then, "simply" discoverable material things, objective facts, or natural kinds, as often assumed?[15] Or are they something else, something more fleeting? After all, species are not just born and gone; they continually unfold. Perhaps studies of extinction and representations of life-forms in cabinets of curiosities need reshuffling on this score. Some scholars argue that it makes more sense to see individual species as historical "processes," as events rather than objects. American evolutionary biologist Olivier Rieppel, for example, insists that species should be presented as open systems with unsharp boundaries, perhaps "compared to super-individuals, such as a colony of corals, that adds to its history as it grows through space and time."[16] A growing number of scholars are exploring the usefulness of the concept of "species continuum," defined as a continuum of reproductive isolation. For Sean Stankowski and Mark Ravinet, such a concept facilitates analyses of population pairs as "snapshots" of particular points in the unfolding of the speciation process.[17]

Alfred Newton would not have been amused by the heat and complexity of modern species talk. In discussions of the naming and classification of species, issues that were hotly debated in his time, Newton was invariably a fundamentalist, faithfully following the Twelfth Edition of Linnaeus's *Systema Naturae*. "With regard to a systemic arrangement of birds," he said, "I am much mistaken if we are not on the verge of the adoption of changes which a short time ago would have admonished the most learned ornithologists, and that in a year or two all the old 'orders' will be entirely broken up and new ones constructed."[18] Subdivisions of "species" irritated him. "I can't see," he once said, "why people can't recognize the existence of breeds or local races without calling them 'sub-species' and giving each a special name. Again . . . if sub-species why not sub-sub-species, any number of subs, in fact?"[19]

Sometimes, when addressing the species problems of his times, Newton was ironic: "I shall not trouble myself about finding out the differences of all these sub-species, I may safely leave details of that kind to those whom it may sub-specially concern."[20] On another occasion, he erupted: "I only wonder I am not driven quite mad and do not dream I am a Gare-fowl's egg about to be involved in a winding sheet of cotton wool and stored away forever in the inmost and most secure compartments of my yellow Lapland coffers."[21]

Beyond Genealogical Trees

Likewise, Newton would not countenance modern criticisms that the "tree"—the standard image of the pedigree of a living object—fails to capture early phases in the history of life in deep time. We should "start with microbes," argues British philosopher of science John Dupré, "rather than with vertebrate animals such as ourselves. The latter are, of course, a very late addition to the biosphere, and a dangerous model for life generally. . . . Only at the other end of evolutionary history does species formation start to look like a general feature of life."[22]

Charles Darwin and Alfred Russel Wallace famously drew "family" trees that showed the advent and trajectory of organisms and species stretching from the distant past to the present. The bifurcation of lineages was shown as tree branches; the extinction of a species was indicated by a twig terminating brusquely. Remarkably, Darwin's most influential "tree of life" drawing made the news again as I was writing this book. Soon after his return from the Galapagos in 1837, from the voyage on which the concept of the evolution of species came to him, Darwin scribbled the first tree diagram in one of his notebooks; it is sometimes called the "I think" diagram, from the first two words on the page. In 2020, Cambridge University Library reported that Darwin's two 1837 notebooks—including the iconic drawing—worth several millions of dollars were missing from the archive, last seen in 2001; the library belatedly reported them stolen, and a worldwide search in partnership with Interpol was launched. As the search stretched over fifteen months, I wondered: Could the thievery of Darwin's drawing have been a theoretical statement or a form of protest? Probably not, I learned in March 2022, when the notebooks were anonymously returned in good condition, in a bright-pink gift bag on the floor of the librarian's office, with a card that read: "Librarian. Happy Easter. X."[23]

Whatever the reason for the theft, scientists and philosophers are indeed protesting the value of Darwin's family trees as a way of conceptualizing life on earth. The tree image is attractive, with a seductive visual appeal, and obviously it can be extremely useful for many things. The concept paved the way for the spectacular successes of genetics in the twentieth and twenty-first centuries—including inspiring scientists studying de-extinction to contemplate bringing a partial imitation of the great auk to life. Genetics also helps to explain why species fail to adapt to rapid environmental change—why, for instance, many seabirds struggle to survive under conditions of overheating, rising sea levels, and acidic waters—adding, in a sense, explanatory power to the theory of natural selection. Had the great auks survived the predation of humans, who rather suddenly gained access to breeding grounds on which the birds had adapted and where they had long been secure, they would now be under another serious threat, much like their nearest kin, including the puffin and the little auk.[24]

Yet tree diagrams have serious drawbacks. In fact, some of the evidence against trees comes from studies of the genetic signatures of birds. One species, the South American hoatzin (*Opisthocomus*, "long-haired bird"), continues to puzzle taxonomists, rebuking straightforward categorization as if the bird belonged to several different lineages. A recent observer, Ben Crair, concludes: "The hoatzin may be more than a missing piece of the evolutionary puzzle. It may be a sphinx with a riddle that many biologists are reluctant to consider: What if the pattern of evolution is not actually a tree?"[25] The lineages of modern birds arrived early on during rapid bursts of evolution and intensive genetic scrambling in the wake of the extinction of the dinosaurs, in the absence of taxonomic enthusiasts.

American writer Jennifer Ackerman adds in her book *The Bird Way* (2020) that genealogical proximity can be deceptive: "One species has evolved one life strategy; that close relative, that phylogenetic neighbor on a twig nearby, does something entirely different."[26] On a family tree, relatedness appears as a simplified geometric abstraction necessarily fixated on connections by birth, leaving no space for connections across lineages or for the passing on of multiple forms of substance and sociality. Life itself appears in a patchy and impoverished form, as a kind of *Wunderkammer*, inundated with endlings.[27]

Another problem is that the tree imagery, in the context of studies of life-forms, tends to reinforce the image of autonomous organisms. Organisms, in fact, are not Robinson Crusoes, all alone in the contemporary life world of the planet and the biosphere. Not only do they live and flourish in the context of others and their shared habitat, but they are also, quite literally, connections themselves. Life comes into existence *because* of connections, American philosopher of science Karen Barad argues.[28] Living beings are not bound by their skin, nor are they, on their own, the direct cause (or the result) of isolated events in the biosphere; they are integrally entwined. Evolution is an interspecies project. Unfortunately, the profound relationality of organisms and species, at the center of modern extinction studies, tends to get lost in the fossil record, as if creatures were devoid of social life and out of geologic time.

What if Darwin—or Alfred Newton for that matter—had begun with microbes, rather than bifurcating branches and trees? What would such an approach mean for the understanding of endlings? The great auk, in the processual sense, was a fleeting moment in the unfolding of life, terminally arrested at a particular point in history by both human and planetary agency. How would one capture its life and extinction in a modern *Wunderkammer*? Perhaps it's time for a new theory of evolution.[29]

Ways of Living and Ways of Talking

The great auk was embedded in a collective habitat, both in the ocean among all its inhabitants and with the other birds nesting on islands and skerries. *Homo sapiens sapiens*, too, is embedded in habitats; we are not immune from the impending extinction of life. After all, our kinsfolk the Neanderthals—who, as Swedish biological anthropologist and Nobel laureate Svante Pääbo demonstrated, contributed substantially to the human genome—vanished around 28,000 years ago. For 100,000 years, they had occupied a "Neanderthal city" of caves on the Iberian peninsula.[30] In their dwellings there, the remains of a hundred and fifty species of birds have been found, including the great auk, indicating that the Neanderthals ate them.[31]

The later cave artists of the distant past did not subscribe to our "species" classification, but they too lived in a world of birds and were surprisingly sharp observers. The Neolithic art of El Tajo in southern Spain, only recently thoroughly studied, captures more than two hundred birds, and no fewer than sixteen species, in amazing detail. Among them is the Great Bustard (*Otis tarda*), the last of which in the area died in 2006.[32]

Around the world, a vast number of bird species, like Iceland's puffin, are now on the brink of extinction.[33] The last sighting of a Hawaiian crow (*Corvus hawaiiensis*), or 'alalā, in the wild was in 2002. Just a handful of 'alalā are now alive in captivity, with the intention of maintaining the species and perhaps even increasing the population by targeted conservation measures. There are lessons to be learned from these last crows. As Wolley and Newton did for the great auk, anthropologist and philosopher Thom van Dooren seeks in his book *Flight Ways* (2014) to attain a deeper understanding of the crows before it is too late—to understand what they stand for, what their loss means to other species, and how they themselves commemorate the members of their species who die.[34] For example, Hawaiian crows are known to systematically avoid places where other crows have been killed. As van Dooren writes:

> But if the death of a single crow signals "here lies danger"—a danger significant enough to avoid a place for years, to alter flight ways and daily foraging routes—then what must the death of a whole species of crow, alongside a host of others at this time, communicate to any sentient and attentive observer? How could these extinctions not announce *our* need to find new flight ways, new modes of living in a fragile and changing world?[35]

Humans must surely think anew, reconsider our "flight ways." It is hard to see how we can allay mass extinction without radical changes to the way we live—without a universal transformation of culture, finance, economics, and communications.

Some species, when they are gone, are more likely to be grieved for than others. But lost loved ones, whether of humans, birds, or beasts, live on in their loss, in the gap they leave behind. A growing literature seeks to render the collective sense of loss and the experience of grief that follows extinction.[36] Humans are not alone in registering this loss. Many species, like the Hawaiian crow, appear to have some awareness of death, and grieving seems to be more common in the animal world than was previously believed. Does this response reflect fear, or grief? Perhaps both. Would this have been the case for the great auk? Probably. Survivors learn from the fate of others; the grieving reaction is a function of lessons learned by previous generations—while also indicating a species' ability to adapt.

For Alfred Newton, as Cowles observes, extinction was "far more serious than cruelty, and he worked hard to disengage the issue of extinction from that of cruelty."[37] Always an astute observer, Newton once remarked that he was sure that two of the great auk eggs he studied—one belonged to him, while the other was at Harvard University, and both had probably been taken on Eldey island in 1840 or 1841—had been laid by the same bird:

> Not only was the character of the spots and markings on each precisely similar, but each possessed at the smaller end a semi-spiral depression, the effect no doubt of a sphincter muscle working upon the shell when in a soft and plastic condition. No one accustomed to eggs could possibly doubt that these . . . were the produce of one and the same parent.[38]

Newton seems to wonder why the bird in question returned each year to her breeding-grounds on Eldey, where hunters took her egg in at least two consecutive years. Significantly, however, he has nothing to say about the auks' feelings of loss. How can the great auk parents have felt, having lost their offspring for two consecutive years? Did they return to Eldey yet again to try a third time? Such questions would be fundamental today, but in Newton's time they were deemed irrelevant: it never occurred to him that a great auk might mourn. That would be, in his terms, naïve sentimentalism.

The great auk would hardly have survived even if the last two birds (or their broken egg) had been spared on Eldey island in 1844. Endangered species are kept "in waiting," for a longer or shorter time—"in the 'death zone,' anticipating the unavoidable," as anthropologist Deborah Bird Rose put it shortly before her own death.[39] The extinction of the great auk may be said to have been impending ever since the massacres near Newfoundland in the 1700s and on Iceland's Great Auk Skerry in the early 1800s. In 1830, the destruction of the skerry, which sank during a subterranean volcanic eruption, added further pressure on the remaining breeding stock. When Vilhjálmur Hákonarson and his crew launched their last successful trip to Eldey, the species had already entered its death zone; in ecological terminology, it was among the "living dead" (too few to breed), along an "extinction spiral" (bound for terminal decline). Extinction never happens directly and quickly, like an execution. Museums and collectors often feel a need to memorialize the last of a species, the endling; yet the last of the great auks—perhaps that same proud bird on its pedestal, that fearlessly demanded my attention in the National History Museum of Denmark—is only one link in a long chain in which organisms and their environment are intertwined, in a reciprocally influential relationship that continues until the very end.

When an organism dies, then, part of its environment dies along with it, affecting a whole network of other organisms of its own and other species. Likewise, the extinction of a single species may trigger the disappearance of hundreds of others.[40] Natural evolution—biological becoming and development—is always relational, mixed up with other creatures and their habitats. Likewise, becoming *endangered* or *extinct* is necessarily both relational and spatial.[41] Organisms and species appear and vanish in the company of others—in a holistic web of life nowadays segregated by human practices, by taxonomies, family trees, species talk, and extinction discourse. As Tim Ingold put it in his poem "On Extinction":

> For here's the question: if all of life is all mixed up,
> Then nothing goes extinct, lest all does.
> There is no extinction, then, without distinction.
> This story of extinction is one that we alone are wont
>   to tell.[42]

Inspired by his 1858 expedition to Iceland with John Wolley, and through careful study of the *Gare-Fowl Books* and other materials bearing witness to the lifeways of the great auk, Alfred Newton made the key intellectual leap of conceiving of extinction as a process—one that could be hastened, or interrupted, by human actions. If only he had expanded this processual concept to the idea of a species itself—relaxing the idea of the *Wunderkammer*, with its immutable slots and universal drawers. Such a perspective of unfolding open systems would be a timely help as we face the complexities of our current age of mass extinction. For Newton's concern with both singularities and collectives, with the mortality of individual great auks as well as of the species and its habitats, was a significant early step in our understanding that what is lost with extinction is not just species and biodiversity but a way of life, natural habitat with an age-old story, a sort of language, even a culture—and the countless lost opportunities of an uncertain collective future that perhaps will never be.

TIMELINE

1854 Philip Henry Gosse publishes his book *The Aquarium: An Unveiling of the Wonders of the Deep Sea.*

1855 Johannes Japetus Steenstrup publishes the first major work on the great auk.

Alfred Russel Wallace develops his theory of natural selection of species.

1858 April 21st: John Wolley and Alfred Newton head off to Iceland from Leith, Scotland, to search for great auks.

May: John Wolley and Alfred Newton hire Geir Zoëga as a guide and translator.

May 8th: John Wolley and Alfred Newton sign a contract with Eiríkur Magnússon to search for great auks in East Iceland.

May 19th: John Wolley and Alfred Newton ride to Kirkjuvogur.

Mid-June: John Wolley and Alfred Newton think that they have established the latest crew for great auk.

July 1st: Charles Darwin and Alfred Russel Wallace announce their theory of natural selection of species.

July 14th: John Wolley and Alfred Newton leave Kirkjuvogur, Iceland, returning to Britain.

1859 Charles Darwin publishes *On the Origin of Species*, on 24 November, establishing "Evolution Day."

Two British amateur archaeologists launch the "time revolution" based on stone tools discovered at Somme in France.

John Wolley dies from brain disease.

1861 Alfred Newton publishes his summary article on the Icelandic expedition.

1864–1907 Alfred Newton publishes *Ootheca Wolleyana.*

1865 Alfred Newton (hesitatingly) pronounces the great auk extinct and launches his notion of unnatural extinction.

1866 Alfred Newton is appointed professor at Magdalene College, Cambridge.

1869 Alfred Newton publishes his article "On the Zoological Aspects of Game Law."

1885 Symington Grieve publishes the first major book about the great auk, *The Great Auk, or Garefowl.*

1900 Gregor Mendel's groundbreaking genetic work from 1866 is independently rediscovered by three scientists.

1907 Alfred Newton dies at Magdalene College, Cambridge.

1941 Margaret Morse Nice publishes her pioneering article "The Role of Territory in Bird Life," a study of birds' biographies.

1999 Errol Fuller publishes *The Great Auk*.
2000 Jeremy Gaskell publishes *Who Killed the Great Auk?*
2013 Henry M. Cowles publishes an article on Alfred Newton's
      works signifying the birth of extinction as a scientific object.
2015 Geneticists meet in Newcastle, England, to discuss the possible
      reanimation of the great auk.
2019 Jessica E. Thomas and colleagues publish an article on the
      genetic diversity of the great auk.
2021 Scientists discover a seabirds' hotspot in the North Atlantic.
      A UN Intergovernmental Panel on Climate Change (IPCC)
      report highlights the risks of mass extinction.
2023 UN delegates reach historic agreement on protecting marine
      diversity.
      200-year anniversary of John Wolley's birth.

ACKNOWLEDGMENTS

Mark Twain once remarked that the more he learned about people, the more he liked his dog. It is tempting to conclude that the same may apply to an anthropologist who turns his attention to birds. Nothing could be further from the truth. What matters, from my perspective, is the interaction between humans and other species, including species of birds. Many bird-lovers may feel there are too many humans in the pages of this book, and it is quite true that Wolley's *Gare-Fowl Books* provide a more detailed account of the great auk than is found here. Other readers may feel that the birds are allocated too much space. In the public mind, as in the various scholarly disciplines, humans and other species belong in separate spheres.

The Last of Its Kind is a heavily expanded and thoroughly revised version of a book that was published in Icelandic as *Fuglinn sem gat ekki flogið* (Mál og menning, 2020). I am grateful to Anna Yates for her skillful English translation of the original Icelandic text and to my agent Michelle Tessler for launching the English edition. My sincere thanks to Princeton University Press editors Alison A. Kalett and Hallie Schaeffer for embracing this project from the start, for their patience and advice, and for pushing me to do my best. Their excellent editorial comments, along with four critical and perceptive anonymous reviews as well as the suggestions of copyeditor Jennifer Harris, significantly helped to reshape the manuscript. My greatest debt is to my friend Nancy Marie Brown, author, scholar, and professional editor. Not only did she comment on my project early on, but also, in her capacity as developmental editor, she applied her extensive scientific knowledge, creative thinking, and brilliant editorial skills to crafting the final narrative. Collectively, this team has "hatched" this book with care and consideration, as if it were a flightless chick in a barren nest on a slippery cliff.

Several people offered guidance and corrections during the writing process. Tim Birkhead, Errol Fuller, Jan Bolding Kristensen, and Kristinn Haukur Skarphéðinsson read the entire manuscript and provided important scientific expertise. Anthropologists and research directors Kristín Erla Harðardóttir and Ragnheiður Hulda Proppé provided vital advice and assistance regarding research applications, funding, and bookkeeping. Without their valuable assistance, this book would not have been possible. My assistant during the writing of the book, librarian and anthropologist Sigurður Örn Guðbjörnsson, deserves thanks for his resourcefulness in gathering a wide range of key resources. Matthías Ægisson drew maps, and Errol Fuller and Kristján Stefánsson helped with finalizing some of the images.

A group of experts, friends, and colleagues have urged me on and smoothed my path in various ways, helped with fieldwork or the gathering of resources, read parts of the book, suggested possible improvements, or drawn my attention to important sources: Jane Acred, Sarah Alison Abel, Hélène Ahlberger Le Deneuff, Ólafur Bjarni Andrésson, Ragnhildur Ágústsdóttir, Michael Brooke, E. Paul Durrenberger, Harri Englund, Marcus Thomas P. Gilbert, Guðmundur Guðmundsson, Elísabet Gunnarsdóttir, Erpur Snær Hansen, Guðrún Haraldsdóttir, Agnar Helgason, Álfheiður Ingadóttir, Nicholas Jones, Dolly Jørgensen, Finn Arne Jørgensen, Petra Tjitske Kalshoven, Örlygur Kristfinnsson, James Laidlaw, Ann Elisabeth Laksfoss Cardozo,

Marianne E. Lien, Bonnie McCay Merritt, Kirstine Møller, Ólöf Nordal, Vífill Oddsson, Ævar Petersen, Guðrún Sigfúsdóttir, Einar Örn Stefánsson, Sverker Sörlin, Jessica E. Thomas, Martin Walsh, Matilda Watson, and Sigríður Þorgeirsdóttir. More than anybody else, Tim Ingold, my supervisor and mentor during my doctoral work in Manchester in the early 1980s, has inspired and influenced my interests and thinking. I express my heartfelt gratitude to all these people. For anything that may be understated, overstated, or wrongly stated, the responsibility is, of course, mine.

My wife, Guðný Sigurbjörg Guðbjörnsdóttir, accompanied me on my research visits and encouraged me throughout, offering reflections on the road. In hindsight, our four grandchildren, Jón Bjarni, Gísli Þór, Saga Rós, and Úlfur Bergmann—sometimes fascinated by stories and pictures of long-gone flightless birds— have probably kept me focused on extinction and on the environmental threats of the modern age. My friend Helgi Bernódusson offered a sober and supportive sounding board through our many discussions during the writing process.

The work on the book was generously supported by the University of Iceland Research Fund. The Cambridge University Library granted me access to the remarkable documents that play a big part in this book, especially John Wolley's *Gare-Fowl Books* and the papers of zoologist Alfred Newton. I thank Domniki Papadimitriou, Johanna Ward, and their colleagues at the library for their helpfulness and collaboration, which was vital for the unfolding of the project. I also thank all the artists, archivists, and publishers who have given me permission to reproduce important material. I am especially thankful for being able to extensively draw upon Alexander Wollaston's *Life of Alfred Newton*, which was a key source of information.

In gathering information for the book, writing, revising, and finalizing the manuscript, I benefited from long stays at the

Icelandic Jón Sigurðsson House (*Jónshús*) in Copenhagen in March 2019; Churchill College, Cambridge, in May of that year; the University of Stavanger, Norway, in March 2022; the writers' residence of Davíðshús in Akureyri, Iceland, in July 2022; and the Swedish Collegium for Advanced Study (SCAS) in Uppsala during the spring of 2023. I am particularly grateful to the Department of Anthropology at Cambridge University, the Greenhouse Environmental Humanities group at Stavanger, and SCAS in Uppsala, for facilitating my long stays. It seemed appropriate, and felt wonderful, to finalize my manuscript and attend to copy editing while at SCAS, in the Linnaean Quarter of Uppsala, where so much species talk must have taken place during and after Linnaeus's time. Thanks to Christina Garsten, principal of SCAS and professor of anthropology, and her colleagues for their hospitality and for hosting me for half a year.

During the writing process, I presented several talks on my work: including a Royal Academy Lecture in the Humanities and Social Sciences in Copenhagen, anthropology seminars in Aarhus and Cambridge, a public talk at the Nordic House in Reykjavík, and seminars in Stavanger and Uppsala. I appreciate the lively discussions at all venues.

Reykjavík and Uppsala, 2023

NOTES

Preface. The Lost Bird

1. Blencowe, *Gone*, ch. 1.
2. Schepelern, *Museum Wormianum*, 284; Steenstrup, "Et Bidrag til Geirfuglens Naturhistorie," 52.
3. Kjartansdóttir, "The Changing Symbolic Meaning of the Extinct Great Auk."
4. Wolley, *Gare-Fowl Books*.
5. Pálsson, "Enskilment at Sea."

Chapter 1. The Road to Extinction

1. Gaskell, *Who Killed the Great Auk?*, 120; Atkinson, *Journal of an Expedition*.
2. Wollaston, *Life of Alfred Newton*, 19.
3. Beer, "Darwin and the Uses of Extinction," 322.
4. Cowles, "A Victorian Extinction."
5. Quotation without a source, in Greene, *The Death of Adam*, 136; see also, Koerner, *Linnaeus*.
6. Müller-Wille and Reeds, "A Translation of Carl Linnaeus's Introduction to *Genera plantarum* (1737)," 565. Emphasis in the original.
7. Ekman, *Kunstkammer*, 5–6.
8. Shubin, "Extinction in Deep Time," 23.
9. Rudwick, *Earth's Deep History*, 108–10.
10. Daston and Galison, *Objectivity*, 89.
11. Torrens, "Mary Anning (1799–1847) of Lyme."
12. Torrens, "Mary Anning (1799–1847) of Lyme."
13. Pyenson, *Spying on Whales*, 125.
14. Emling, "Mary Anning," 6 and 7.
15. See Purchell and Gould, *Finders, Keepers*, 100.
16. Gamble, *Making Deep History*.
17. See Smith, *Alfred Russel Wallace*, 222.
18. Darwin, *On the Origin of Species by Means of Natural Selection*, 154.

19. Barrow, *Nature's Ghosts*.

20. Darwin, *On the Origin of Species*, 342.

21. Delisle, *Charles Darwin's Incomplete Revolution*.

22. Fuller, *The Great Auk*, 42.

23. Bengtson, "Breeding Ecology and Extinction of the Great Auk (*Pinguinus impennis*)," 9.

24. Thomas et al., "Demographic Reconstruction from Ancient DNA Supports Rapid Extinction of the Great Auk"; Evans, "From Healthy to Extinct in 350 Years."

25. Quotation in Gaskell, *Who Killed the Great Auk?*, 125.

26. Thorsen, *Dyrenes by*.

27. Steenstrup, "Et Bidrag til Geirfuglens Naturhistorie."

28. Steenstrup, "Et Bidrag til Geirfuglens Naturhistorie," 65.

29. *Naturhistorisk tidskrift*, 207; Grieve, *The Great Auk, or Garefowl*, 2.

30. Cowles, "A Victorian Extinction," 696.

31. Birkhead and Gallivan, "Alfred Newton's Contributions to Ornithology."

32. Pálsson, *The Human Age*.

33. Sklair (ed.), *The Anthropocene in Global Media*; Rose et al. (eds.), *Extinction Studies*.

34. Zimmer, "Birds Are Vanishing from North America."

35. Weston, "Half of World's Bird Species in Decline as Destruction of Avian Life Intensifies."

36. Whitehouse, "Listening to Birds in the Anthropocene."

37. Newton, "The Gare-Fowl and Its Historians," 477.

38. Letter from Plesner to Rowley, 19 January 1873. *Gare-Fowl Books*, book 4, 83–86.

39. Jørgensen, "Endling"; Pyne, *Endling*.

40. Kingsley, *The Water-Babies*, 289.

41. Rose et al. (eds.), *Extinction Studies*.

42. See Cowles, "A Victorian Extinction," and Kolbert, *The Sixth Extinction*, ch. III.

43. Cowles, "A Victorian Extinction."

44. Newton, "The Gare-Fowl and Its Historians."

Chapter 2. A Very Victorian Quest

1. Wollaston, *Life of Alfred Newton*, 45–46.

2. Cole, "Blown Out"; Walters, "Uses of Egg Collections."

3. Bourne, "The Story of the Great Auk *Pinguinis impennis*," 272.

4. Newton, "Memoir of the Late John Wolley," 172.

5. Quammen, *The Song of the Dodo*.

6. See Wollaston, *Life of Alfred Newton*, 16.

7. Danell, "Äggherren och hans runsten i Muoniovaara."

8. Palmbo, "Äggherren och runristningen i Muoniovaara"; Nordén, *Sällsamheter i Tornedalen.*

9. Michelle Jennen, e-mail to the author, 17 February 2023.

10. See Wollaston, *Life of Alfred Newton,* 70; Wolley and Newton, *Ootheca Wolleyana.*

11. Birkhead, Axon, and Middleton, "Restoration of Two Great Auk (*Pinguinus impennis*) Eggs."

12. Wollaston, *Life of Alfred Newton,* 2.

13. Pálsson, *The Man Who Stole Himself.*

14. Cunich et al., *A History of Magdalene College Cambridge, 1428–1988.*

15. Legacies of British Slave-Ownership, website.

16. Kingsley, *The Water-Babies.*

17. Newton and Newton, "Observations on the Birds of St. Croix, West Indies," 140.

18. Kjartansdóttir, "The Changing Symbolic Meaning of the Extinct Great Auk"; Schepelern, *Museum Wormianum,* 284; Steenstrup, "Et Bidrag til Geirfuglens Naturhistorie," 52.

19. de Lozoya et al., "A Great Auk for the Sun King."

20. Thisted, *Således skriver jeg, Aron,* 478; Meldgaard, "The Great Auk, *Pinguinus impennis* (L.) in Greenland," 152.

21. Clottes et al., *Cosquer Redécouvert.*

22. Eastham and Eastham, "Palaeolithic Images and the Great Auk," 1023.

23. See Alves and Albuquerqu (eds.), *Ethnozoology.*

24. Kolbert, *The Sixth Extinction,* ch. III.

25. Olson and Lund, "Whalers and Woggins."

26. Grieve, *The Great Auk, or Garefowl,* 131.

27. Serjeantson, "The Great Auk and the Gannet"; Steenstrup, "Et bidrag til geirfuglens naturhistorie."

28. Braun, "Representations of Birds in the Eurasian Upper Palaeolithic Ice Age Art," 16; Finlayson, *The Smart Neanderthal,* ch. 9.

29. Speck, *Bird Lore of the Northern Indians,* 353.

30. Pope, "Early Migratory Fishermen and Newfoundland's Seabird Colonies," 62.

31. See Pope, "Early Migratory Fishermen and Newfoundland's Seabird Colonies," 64; see also Birkhead, *Great Auk Islands.*

32. Newton, "Abstract of Mr. J. Wolley's Researches in Iceland Respecting the Gare-Fowl or Great Auk (*Alca impennis,* Linn)," 385–86.

Chapter 3. An Awkward Trip to Iceland

1. Newton, "Abstract of Mr. J. Wolley's Researches," 377.

2. Letter of Wolley to Newton, 24 March 1858.

3. Letter of Wolley to Newton, 1 April 1858.

4. Letter of Wolley to Newton 5 April 1852.

5. Maurer, *Íslandsferð 1858*, 10.

6. Letter of Alfred Newton to Edward Newton, 2 May 1858. See Wollaston, *Life of Alfred Newton*, 29.

7. Wollaston, *Life of Alfred Newton*, 28.

8. Wolley, *Gare-Fowl Books*, book 1, 3–4.

9. Wollaston, *Life of Alfred Newton*, 30.

10. See Karlsson, *A Brief History of Iceland*.

11. Helgason, "Reykjavíkurkaupmenn á æskuárum mínum," 4.

12. Baring-Gould and Newton, *Iceland*, 26–27.

13. Óskarsson, *Saga Reykjavíkur*, 250.

14. Daston and Galison, *Objectivity*.

15. Wolley, *Gare-Fowl Books*.

16. Faber, "Beyträge zur arctischen Zoologie."

17. Jónsson, *Þjóðhættir og ævisögur frá 19. öld*, 151.

18. Sæmundsson, "Fiskirannsóknir 1911–1912," 22.

19. Wollaston, *Life of Alfred Newton*, 31.

20. Wolley, *Gare-Fowl Books*, book 3, no page numbers.

21. Wollaston, *Life of Alfred Newton*, 60.

22. Wollaston, *Life of Alfred Newton*, 33.

23. Wollaston, *Life of Alfred Newton*, 31.

24. Newton, "Abstract of Mr. J. Wolley's Researches," 399.

25. Newton, "On the Zoological Aspect of Game Laws," 107–8.

26. Heaton, "We're Still Fighting Alfred's Battles 150 Years On."

27. Birkhead and Gallivan, "Alfred Newton's Contribution to Ornithology," 13; Cowles, "A Victorian Extinction," 710.

28. Sea Birds Preservation Act, 1869, website.

29. Bargheer, "The Sociology of Morality as Ecology of Mind," 79–80.

30. Birkhead, *Birds and Us*, 308; Letter of A. Newton to H. F. Barnes of 27 February 1872, quoted in Cowles, "A Victorian Extinction," 711.

31. Birkhead and Gallivan, "Alfred Newton's Contributions to Ornithology," 899.

32. Birkhead, *Birds and Us*, 222.

Chapter 4. Arriving

1. Newton, "Memoir of the Late John Wolley," 13.

2. Wollaston, *Life of Alfred Newton*, 34.

3. Wollaston, *Life of Alfred Newton*, 35.

4. Mynott, *Birds in the Ancient World*.

5. Þór, *Hafnir á Reykjanesi*; Thorarensen, "Um Hafnir í gamla daga."

6. Magnússon and Vídalín, *Jarðabók*, 29–32.

7. Alfred Newton's Papers, unpublished sources.

8. Wolley, *Gare-Fowl Books*, book 1, 39.

9. Wollaston, *Life of Alfred Newton*, 36.

10. Wolley, *Gare-Fowl Books*, book 1, 121.

11. Letter from Müller to Wolley, 28 May 1858.

12. Thorarensen, "Suðurnesjaannáll 1858."

13. Aho, "Með Ísland á heilanum."

14. Agnarsdóttir, "Sir Joseph Banks and the Exploration of Iceland."

15. Quoted in Duyker, *Nature's Argonaut*, 252.

16. Newton, "Appendix A," 399–421.

17. Guðbjörnsson, "Sundreglur prófessors Nachtegalls," unpublished manuscript.

18. Gosse, *The Aquarium*, iii.

19. Granata, *The Victorian Aquarium*, 7.

20. Granata, *The Victorian Aquarium*, 198; original emphasis.

21. Pálsson, "The Birth of the Aquarium"; Brunner, *The Ocean at Home*.

22. Ashworth, "HMS Challenger," website.

23. Montgomery, *The Hummingbirds' Gift*.

24. Gosse's letter to Newton, 5 January 1857; also see Gosse's letter of 9 January.

25. Rose, *Shimmer*.

26. Quoted in Daston and Galison, *Objectivity*, 64.

Chapter 5. Rebooting

1. Kolbert, *The Sixth Extinction*, 62–63.

2. Kolbert, *The Sixth Extinction*, 64.

3. *Morgunblaðið*, "Gríðarlega mikil plastmengun í Eldey," 2.

4. Wollaston, *Life of Alfred Newton*, 38.

5. Newton, "Abstract of Mr. J. Wolley's Researches," 380, footnote.

6. *Útfararminning dannebrogsmanns Vilhjálms Hákonarsonar*, 3–4.

7. Alfred Newton's Papers, 111 (ii).

8. Wollaston, *Life of Alfred Newton*, 32–33.

9. Ólafsson and Pálsson, *Rejse igiennem Island*.

10. Houston et al., "How Did the Great Auk Raise Its Young?" 95; Birkhead, "The Chick-Rearing Period of the Great Auk"; Wolley, *Gare-Fowl Books*, book 1, 42, 59.

11. Einarsson, *Saga Eiríks Magnússonar í Cambridge*, 15.

12. Newton, "Abstract of Mr. J. Wolley's Researches," 395.

13. Newton, "Abstract of Mr. J. Wolley's Researches," 396.

14. See Spray, "The Terrible *Njorl's Saga*," 108.

15. Lambert, "From Exploitation to Extinction, to Environmental Icon," 28.

16. Cowles, "A Victorian Extinction."

17. Endersby, "Sympathetic Science."

18. Cowles, "A Victorian Extinction."

19. Newton, "Address to the Department of Botany and Zoology," 125.

20. Quotation in Jónasson et al., *Híbýli fátæktar*, 32.

21. Pálsson and Durrenberger, "Icelandic Foremen and Skippers"; Pálsson, *Coastal Economies, Cultural Accounts*.

22. Wolley, *Gare-Fowl Books*, book 1, 106.

23. Wolley, *Gare-Fowl Books*, book 1, 68.

24. Gaskell, *Who Killed the Great Auk?*, 127.

25. Guðmundsson, *Íslenzkir athafnamenn I*, 75.

26. Guðmundsson, *Íslenzkir athafnamenn I*, 71.

27. Browne, *The Land of Thor*, 440

28. Aho, "Með Ísland á heilanum."

29. Einarsson, "Það var hin mesta hættuför," 22–23, 51.

30. DeLoughrey, *Allegories of the Anthropocene*.

31. Grimbert, *Le Dernier des siens*.

32. Lund et al., "Puffin Love."

33. Koons, "Becoming Avian in the Anthropocene," 108.

34. Greengrass, *An Account of the Decline of the Great Auk*, 10.

35. Marguilies, "A Political Ecology of Desire"; Barnett, *Mourning in the Anthropocene*.

36. Cherry-Garrard, *The Worst Journey in the World*, 396.

37. Letter from Newton to T. Southwell, Esq., 30 August 1858; Wollaston, *Life of Alfred Newton*, 40.

38. Headley, "Introduction"; Pyne, "What a New Translation of *Beowulf* Says about Extinction."

Chapter 6. "Proud Birds, with Their Heads High"

1. Audubon, *Birds of America*; Rhodes, *John James Audubon*.

2. Wollaston, *Life of Alfred Newton*, 7.

3. Newton, "The Gare-Fowl and Its Historians," 486.

4. Daston and Galison, *Objectivity*, 79.

5. Kalshoven, "Piecing Together the Extinct Great Auk," 154.

6. Jónsson, *Dýrafræði*, vol. 2, 71.

7. Wolley, *Gare-Fowl Books*, book 1, 35.

8. Rev. Sívertsen, in Wolley, *Gare-Fowl Books*, book 1, 86.

9. Wolley, *Gare-Fowl Books*, book 1, 91.

10. Wolley, *Gare-Fowl Books*, book 1, 93.

11. See Morris and Duggins, "The Avian Blink."

12. Wolley, *Gare-Fowl Books*, book 2, 128.

13. E-mail from Wolfgang Müller, 7 April 2019.

14. Lorimer, "On Auks and Awkwardness"; see also Oxford English Dictionary, website.

15. Müller, *Séance Vocibus Avium*, website.

16. Wolley, *Gare-Fowl Books*, book 1, 26.

17. Despret, *Living as a Bird*.

18. Hoare, "Under the Skin of the Ocean, There's a Super-Loud Fishcotheque Going On"; Richardson, *Where the Seals Sing*.

19. Davies et al., "Multispecies Tracking Reveals a Major Seabird Hotspot in the North Atlantic."

20. Thomas et al., "Demographic Reconstruction from Ancient DNA Supports Rapid Extinction of the Great Auk."

21. Beer, "Darwin and the Uses of Extinction," 326.

22. Kalshoven, "Piecing Together the Extinct Great Auk," 157; Haraway, "Teddy Bear Patriarchy"; Ackerman, *The Bird Way*, 19.

23. Clottes et al., *Cosquer Redécouvert*, 130; Reisz, "Great Auks and Seal-headed Men."

24. Kalshoven, "Piecing Together the Extinct Great Auk," 165–66.

25. Bourne, "The Story of the Great Auk," 272.

26. Wolley, *Gare-Fowl Books*, book 2, 258–59.

27. Martin, *A Late Voyage to St. Kilda*, 48.

28. Birkhead et al., "The Great Auk (*Pinguinus impennis*) Had Two Brood Patches," 4.

Chapter 7. Questions In, and Out of, Time

1. Wolley, *Gare-Fowl Books*, book 2, 130.

2. See Wolley, *Gare-Fowl Books*, book 4, 355.

3. de Lozoya et al., "A Great Auk for the Sun King," 50.

4. Lien and Pálsson, "Ethnography beyond the Human"; Lestel et al., "The Phenomenology of Animal Life"; Hill, "Archaeology and Animal Persons."

5. Ingold, *Imagining for Real*, 306.

6. Haraway, *When Species Meet*; Pálsson, "Biosocial Relations of Production"; Abrell, *Saving Animals*.

7. Posnett, "The Weird Magic of Eiderdown."

8. Matza, "Why US Bird Attacks on Humans Are on Rise," website.

9. Krebber and Roscher (eds.), *Animal Biography*, 2.

10. Birkhead, *Birds and Us*, 284.

11. Despret, "Inhabiting the Phonocene with Birds," 256; see also Despret, *Living as a Bird*.

12. Despret, *The Dance of the Arabian Babbler*, 3.

13. Aamodt et al., "Birdsong as a Window into Language Origins and Evolutionary Neuroscience"; Ackerman, *The Bird Way*.

14. Tsing, "The Sociality of Birds," 29.

15. von Uexküll, *A Foray into the Worlds of Animals and Humans*.

16. Schroer, "Jacob von Uexküll."

17. See Schroer, "Jakob von Uexküll," 145–46.

18. Nice, "The Role of Territory in Bird Life."

19. Gosse, *The Aquarium*, iv–v.

20. Despret, *Living as a Bird*, 116.

21. Hastrup, *Nature and Policy in Iceland 1400–1800*, 245.

22. Wolley, *Gare-Fowl Books*, book 1, 120.

23. Wolley, *Gare-Fowl Books*, book 1, 34–36.

24. Falk, *Violence and Risk in Medieval Iceland*, 225.

25. Wolley, *Gare-Fowl Books*, book 1, 120.

26. See Wolley, *Gare-Fowl Books*, book 2, 229.

27. Wolley, *Gare-Fowl Books*, book 1, 60.

28. Newton, "Abstract of Mr. J. Wolley's Researches," 380, 390.

29. Kristjánsson, "Geirfugl," 263–65.

30. Ólafsson and Pálsson, *Ferðabók Eggerts Ólafssonar og Bjarna Pálssonar*, 155.

31. Rev. Sívertsen, see Wolley, *Gare-Fowl Books*, book 1, 85.

32. Newton, "Abstract of Mr. J. Wolley's Researches," 382.

33. Wolley, *Gare-Fowl Books*, book 2, 252.

34. Þórarinsson, "Neðansjávargos við Ísland," 65.

35. Newton, "Abstract of Mr. J. Wolley's Researches," 382.

36. Wolley, *Gare-Fowl Books*, book 1.

37. Árnason, *Íslenskar þjóðsögur og ævintýri*, 82–88.

38. Brown, *Looking for the Hidden Folk*; Raffles, *The Book of Unconformities*, 101–22.

39. Guðmundsson, "Lýsing á Höfnum," 57–58.

40. Wolley and Newton, *Ootheca Wolleyana*, xxvi.

41. Wolley, *Gare-Fowl Books*, book 1, 102.

Chapter 8. The Latest Successful Trip

1. Málfarsbankinn, website.
2. Thorarensen, "Suðurnesjaannáll 1844."
3. Wolley, *Gare-Fowl Books*, book 1, 113; Book of Icelanders, website.
4. Wolley, *Gare-Fowl Books*, book 1, 74.
5. Willson, *Seawomen of Iceland*.
6. Manuscript of Jón Bjarnason (about 1791–1861), unpublished source.
7. Lambert, "From Exploitation to Extinction, to Environmental Icon," 22.
8. Wolley, *Gare-Fowl Books*, book 1, 110.
9. Wolley, *Gare-Fowl Books*, book 1, 41.
10. Wolley, *Gare-Fowl Books*, book 1, 41.
11. Newton, "Abstract of Mr. J. Wolley's Researches," 375.
12. Newton, "Abstract of Mr. J. Wolley's Researches."
13. Gaskell, *Who Killed the Great Auk?*, 130–31.
14. Wolley, *Gare-Fowl Books*, book 1, 59.
15. Newton, "Abstract of Mr. J. Wolley's Researches," 399.

Chapter 9. The Human Drama

1. Fuller, *The Great Auk*.
2. *Extinct Birds*; *Dodo*; *The Passenger Pigeon*.
3. Attenborough and Fuller, *Drawn from Paradise*.
4. Quotation in Greenlaw, *Questions on Travel*, xviii.
5. Wollaston, *Life of Alfred Newton*, 39.
6. Letter from Hjaltalín to Wolley, 4 December 1858.
7. Letter from Zoëga to Wolley, 2 June 1859.
8. Newton, "Abstract of Mr. J. Wolley's Researches."
9. Newton, "Abstract of Mr. J. Wolley's Researches."
10. *Frjáls þjóð*, "Brúðarrán á Suðurnesjum," 3.
11. Nicolson, *The Seabird's Cry*, 287.
12. Letter from Wolley to Newton, 3 September 1858.
13. Newton, "Memoir of the Late John Wolley."
14. Letter from Wolley to Newton, 27 July 1859.
15. Historicracing.com, website; Ancestry.ca, website.
16. Newton, "Memoir of the Late John Wolley."
17. Wolley and Newton, *Ootheca Wolleyana*.
18. Daston and Galison, *Objectivity*, 26.
19. Le Quellec, *La caverne originelle*; Clottes, *What Is Paleolithic Art?*; Porr, "Rock Art as Art."

20. Wawn, "Fast er drukkið og fátt lært."

21. Einarsson, *Saga Eiríks Magnússonar í Cambridge*, 15.

22. Wolley, *Gare-Fowl Books*, book 4, cv.

23. Newton, "On Existing Remains of the Gare-Fowl (*Alca impennis*)," 257.

24. Grieve, *The Great Auk, or Garefowl*, 2.

25. Newton, "The Great Auk, or Garefowl (Alca impennis, Linn), Its History, Archaeology, and Remains," 546.

26. Fuller, *The Great Auk*, 384.

27. Newton, "The Gare-Fowl and Its Historians," 477.

28. Wood, "The History of Bird Banding"; Weidensaul, *A World on the Wing*.

29. Icelandic ornithologist Finnur Guðmundsson (1909–79); see Pálsson, "Vega-bréf fuglanna."

30. Newton, "On the Possibility of Taking an Ornithological Census," 193.

31. Newton, "On the Possibility of Taking an Ornithological Census," 191.

32. Birkhead, *Birds and Us*, 381.

33. Buffon, *The Epochs of Nature*.

34. Cowles, "A Victorian Extinction," 696.

35. Owen, "On the Extinction of Species," 56, Appendix A.

36. Sepkoski, *Catastrophic Thinking*, 49.

Chapter 10. Newtonian Extinction

1. Letter to H. B. Tristram, 2 February 1888; Wollaston, *Life of Alfred Newton*, 117–18.

2. Barrow, *Nature's Ghosts*.

3. Quotation in Wollaston, *Life of Alfred Newton*, 112.

4. Letter to Tristram, 2 February 1888; Wollaston, *Life of Alfred Newton*, 118.

5. Letter from Wolley to Newton, 8 October 1858.

6. Feeley-Harnik, "The Geography of Descent."

7. Letter from Newton to Darwin, 21 January 1867. Darwin Correspondence Project, website.

8. Quoted in Johnson, "Charles Darwin, Richard Owen, and Natural Selection," 51.

9. Lomolina, "Wallace at the Foundations of Biogeography and the Frontiers of Conservation Biology," 351.

10. Letter of Wallace to Newton, 19 February 1865.

11. Newton, "Cuckoo's Eggs."

12. Letter of Darwin to Newton, 8 October 1862; Wollaston, *Life of Alfred Newton*, 124–25.

13. Beer, "Darwin and the Uses of Extinction," 323.

14. Letter of Darwin to Newton, 29 October 1865.

15. Cowles, "A Victorian Extinction," 704.

16. Letter to Edward Newton, 25 July 1860; Wollaston, *Life of Alfred Newton*, 118–20.

17. Darwin, *The Descent of Man*.

18. Cowles, "A Victorian Extinction," 703.

19. Fuentes, "'The Descent of Man,' 150 Years On"; Ingold, *Evolution and Social Life*.

20. Letter to Newton, 8 October 1862; Wollaston, *Life of Alfred Newton*, 123.

21. Johnson, "Charles Darwin, Richard Owen, and Natural Selection."

22. Darwin, *On the Origin of Species*, ch. 1, p. 76.

23. Müller-Wille and Rheinberger (eds.), *Heredity Produced*.

24. Galton, "Did Darwin Read Mendel?" 588; Tudge, *In Mendel's Footnotes*.

25. Smocovitis, *Unifying Biology*.

26. Wollaston, *Life of Alfred Newton*, 104.

27. Cowles, "A Victorian Extinction."

28. Owen, *Palæontology*, 399; my emphasis.

29. Owen, *Palæontology*, 400.

30. Newton, "Abstract of Mr. J. Wolley's Researches," 397.

31. Newton, "Abstract of Mr. J. Wolley's Researches," 397.

32. Newton, "Abstract of Mr. J. Wolley's Researches," 398.

33. Newton, "The Gare-Fowl and Its Historians," 487.

34. Newton, quoted in Cowles, "A Victorian Extinction," 714.

35. Daston and Galison, *Objectivity*, 17.

36. Cowles, "A Victorian Extinction," 697.

37. Newton, "Testimony before the Select Committee on Wild Birds Protection," unpublished source, 34.

38. Cowles, "A Victorian Extinction."

39. Thomas, "Allegorizing Extinction."

40. Wulf, *The Invention of Nature*.

41. Walls, *Henry David Thoreau*, xvi.

42. Marsh, *Man and Nature*, 84.

43. Worster (ed.), *The Ends of the Earth*, 4.

44. Worster (ed.), *The Ends of the Earth*, 17.

45. Ehrlich and Ehrlich, *Extinction*, 225.

46. IPCC/United Nations Report, website; Watts, "Biodiversity Crisis Is About to Put Humans at Risk."

47. Chernela, "A Species Apart," 27.

48. Bennett, *The Ecological Transition*.

49. Haraway, *Staying with the Trouble*, 40.

50. Rose, "Double Death."

51. Connolly, *Facing the Planetary*; Latour and Weibel, *Critical Zones*.

52. Rutz et al., "Covid-19 Lockdown Allows Researchers to Quantify the Effects of Human Activity on Wildlife."

53. Meloni et al., "Bodies of the Anthropocene"; Ingold and Pálsson (eds.), *Biosocial Becomings*; Lock and Pálsson, *Can Science Resolve the Nature/Nurture Debate?*

54. Owen, "On the Extinction of Species."

55. Arendt, *The Human Condition*, 47; my emphasis.

56. Newton, *A Dictionary of Birds*, 215.

57. Barad, "Posthumanist Performativity."

58. Kolbert, *The Sixth Extinction*; Hannah, *Extinctions*; Nijhuis, *Beloved Beasts*.

59. Hannah, *Extinctions*.

Chapter 11. The Last Great Auk

1. Nielsen, "Geirfugl, sæörn, og fálki."

2. Ketilsson, "Síðustu geirfuglarnir," 2.

3. Ketilsson, "Síðustu geirfuglarnir," 2.

4. Nielsen, "Síðustu geirfuglarnir."

5. Gaskell, *Who Killed the Great Auk?*, 129; see also Barrow, *Nature's Ghosts*, 62.

6. "The Extinction of the Great Auk." Audubon, website.

7. Nielsen, "Síðustu geirfuglarnir."

8. Bragadóttir, "Fugla- og náttúruvernd Nielsens kaupmanns," 330.

9. Ketilsson, "Geirfugl, örn, og valur."

10. Ísaksson, *Úr sögu geirfuglsins á Íslandi*, 35.

11. Ólason, *Íslenskar æviskrár frá landnámstímum til ársloka 1940*, 356.

12. The Glyptodon, website.

13. Interview with Oddsson, Reykjavík, 1 April 2019.

14. Andrésson, "Enn vil ég vera Ketill."

15. Wolley, *Gare-Fowl Books*, book 3, 89.

16. Hagalín, *Saga Eldeyjar-Hjalta*, 246–47.

17. Hagalín, *Saga Eldeyjar-Hjalta*, 248.

18. Hagalín, *Saga Eldeyjar-Hjalta*, 253; see also Þórarinsson, "Eldeyjarfarir fyrir 60 árum," 173–88.

19. Hagalín, *Saga Eldeyjar-Hjalta*, 254.

20. Fuller, *The Great Auk*, 404–13; Gaskell, *Who Killed the Great Auk?*, ch. 13; Ísaksson, *Úr sögu geirfuglsins á Íslandi*, 33.

21. Bird Life International, website.

22. Kalshoven, "Piecing Together the Extinct Great Auk," 152.

23. Letter from Plesner to Rowley, 19 January 1873; Wolley, *Gare-Fowl Books*, book 4, 83–86.

24. Unsöld, "Two and a Half Auks."

25. Fuller, *The Great Auk*, 85.

26. See Kalshoven, "Piecing Together the Extinct Great Auk," 165.

27. Thomas et al., "An 'Aukward' Tale"; Thomas, *Evolution and Extinction of the Great Auk.*

28. Wollaston, *Life of Alfred Newton*, 324.

29. Wollaston, *Life of Alfred Newton*, 234.

30. Wollaston, *Life of Alfred Newton*, 270.

31. Wollaston, *Life of Alfred Newton*, 105.

32. Letter from Newton to J. A. Harvie-Brown, 5 April 1905. See Wollaston, *Life of Alfred Newton*, 291.

33. Wollaston, *Life of Alfred Newton*, 273.

34. Birkhead, *Birds and Us*, 301, 302.

35. Birkhead and Gallivan, "Alfred Newton's Contributions to Ornithology," 899.

36. Birkhead, *Birds and Us*, 303–4.

37. Nijhuis, *Beloved Beasts*, 35.

38. Wilkins, "Ten Extinct Animals Have Been Rediscovered," website.

39. Darwin, *On the Origin of Species*, 319.

40. Monbiot, *Feral*; Barkham, "It's Very Easy to Save a Species."

41. Brand, "Resurrecting the Great Auk," website.

42. Interview with Thomas Gilbert, 22 March 2019.

43. Richmond et al., "The Potential and Pitfalls of De-extinction."

44. Brogaard, "Species as Individuals."

45. Fletcher, *De-extinction and the Genomics Revolution*; Barkham, "It's Very Easy to Save a Species."

46. Scales, "Is It Time to Begin Rewilding the Oceans."

47. van Dooren, *A World in a Shell*, 9.

Afterword. On the Brink

1. Lewontin, "Gene, Organism, and Environment," 63.

2. Despret, *Living as a Bird*, 102; original emphasis.

3. Costa, "Wallace, Darwin, and Natural Selection," 107.

4. Ehrlich and Ehrlich, *Extinction*, 17.

5. Barrowclough et al., "How Many Kinds of Birds Are There and Why Does It Matter?"

6. Crair, "The Bizarre Bird That's Breaking the Tree of life."

7. Wilson, "Continuing after Species," 345.

8. Mishler, "Ecology, Evolution, and Systematics in a Post-Species World," 188.

9. Amitani, "Is the Species Concept *That* Important?"; Rheinberger, "Gene Concepts."

10. Atran, "Folk Biology and the Anthropology of Science"; Wilkins and Zachos, *Species Problems and Beyond*; Sodikoff, *The Anthropology of Extinction*.

11. Barnett, *Mourning in the Anthropocene*, ch. 2; Butler, *Precarious Life*.

12. Athreya and Hopkins, "Conceptual Issues in Hominin Taxonomy."

13. Eastham and Eastham, "Palaeolithic Images and the Great Auk," 1024; emphasis is mine.

14. Sigari et al., "Birds and Bovids," 1399.

15. Daston and Galison, *Objectivity*, ch. 2.

16. Rieppel, "Species as a Process," 45.

17. Stankowski and Ravinet, "Defining the Speciation Continuum," 1257.

18. Letter from Newton to Mrs. Strickland, 12 February 1867. See Wollaston, *Life of Alfred Newton*, 214–15.

19. Wollaston, *Life of Alfred Newton*, 218–19.

20. Letter from Newton to J. A. Harvie-Brown, 20 July 1904. See Wollaston, *Life of Alfred Newton*, 218.

21. Letter of Newton to Tristram on 11 July 1863. See Wollaston, *Life of Alfred Newton*, 76–77.

22. Dupré, "(Some) Species Are Processes," 281; see also Dupré, *Processes of Life*.

23. Parkin, "Darwin's Lost Treasure, Found"; Roberts, "Missing Darwin's Notebooks Returned to Cambridge University Library," website; Pietsch, *Trees of Life*, 92.

24. Jakubas et al., "A Quiet Extirpation of the Breeding Little Auk *Alle alle* Population in the Shadow of the Famous Cousin Extermination."

25. Crair, "The Bizarre Bird That's Breaking the Tree of Life," 3.

26. Ackerman, *The Bird Way*, 318.

27. Ingold, *The Rise and Fall of Generation Now*, ch. 5.

28. Barad, "Posthumanist Performativity," 821.

29. Buranyi, "Do We Need a New Theory of Evolution?"

30. Papagianni and Morse, *The Neanderthals Rediscovered*, 138–43.

31. Hogenboom, "How Did the Last Neanderthals Live?"; Birkhead, *Birds and Us*, ch. 1.

32. Birkhead, *Birds and Us*, 8, 12.

33. Weidensaul, *A World on the Wing*.

34. van Dooren, *Flight Ways*, ch. 5.

35. van Dooren, *Flight Ways*, 142; original emphasis.

36. Rose et al., *Extinction Studies*; Jørgensen, *Recovering Lost Species in the Modern Age*.

37. Cowles, "A Victorian Extinction," 707.

38. Wolley and Newton, *Ootheca Wolleyana*, vol. 2, 376.

39. Rose, "Slowly—Writing into the Anthropocene"; Rose, *Shimmer*; see also Abrell, *Saving Animals*.

40. Ackerman, *The Bird Way*, 324; van Dooren, *A World in a Shell*.

41. Garlick and Symons, "Geographies of Extinction."

42. Ingold, "On Extinction."

BIBLIOGRAPHY

Published Material

Aamodt, Catlina M., Madza Farias-Virgens, and Stephanie A. White. "Birdsong as a Window into Language Origins and Evolutionary Neuroscience." *Philosophical Transactions of the Royal Society B* 375 (2019). https://doi.org/10.1098/rstb.2019.0060.

Abrell, Elan. *Saving Animals: Multispecies Ecologies of Rescue and Care.* Minneapolis: University of Minnesota Press, 2021.

Ackerman, Jennifer. *The Bird Way: A New Look at How Birds Talk, Work, Play, Parent, and Think.* New York: Penguin Books, 2021.

Agnarsdóttir, Anna. "Sir Joseph Banks and the Exploration of Iceland." In *Sir Joseph Banks: A Global Perspective*, edited by R.E.R. Banks, B. Elliott, J. G. Hawkes, D. King-Hele, and G. L. Lucas, 31–48. Kew: Royal Botanic Gardens, 1994.

Aho, Gary. "Með Ísland á heilanum: Íslandsbækur breskra ferðalanga 1772–1897." *Skírnir* 167 (1993): 205–58.

Alves, Romulo Romeu Nobrega, and Ulysses Paulino Albuquerqu (eds.). *Ethnozoology: Animals in Our Lives.* London: Academic Press, 2017.

Amitani, Yuichi. "Is the Species Concept *That* Important?" In *Species Problems and Beyond: Contemporary Issues in Philosophy and Practice*, edited by John Wilkins, Franz Zachos, and Igor Pavlinov, 39–63. Boca Raton, FL: CRC Press, 2022.

Andrésson, Ólafur Bjarni. "Enn vil ég vera Ketill." *Morgunblaðið*, 18 August 2020.

Arendt, Hannah. *The Human Condition.* Chicago: University of Chicago Press, 1958.

Árnason, Jón. *Íslenzkar þjóðsögur og æfintýri*, vol. 1. Leipzig: J. C. Hinrichs, 1862.

———. *Íslenzkar þjóðsögur og æfintýri*, vol. 1, edited by Árni Böðvarsson and Bjarni Vilhálmsson. Reykjavík: Þjóðsaga, 1954.

Arnold, Michael L. *Evolution through Genetic Exchange.* Oxford: Oxford University Press, 2006.

Athreya, Sheela, and Allison Hopkins. "Conceptual Issues in Hominin Taxonomy: *Homo heidelbergensis* and Ethnobiological Reframing of Species." *Yearbook of Physical Anthropology* 175 Suppl. 72 (2021): 4–26.

Atkinson, George Clayton. *Journal of an Expedition to the Faroe and Westman Islands and Iceland 1833*. Newcastle upon Tyne: Bewick-Beaufort Press, 1989 [1833].

Atran, Scott. "Folk Biology and the Anthropology of Science: Cognitive Universals and the Cultural Particulars." *Behavioral and Brain Sciences* 21, no. 4 (1998): 547–609. doi: 10.1017/s0140525x98001277.

Attenborough, Sir David, and Errol Fuller. *Drawn from Paradise: The Discovery, Art and Natural History of the Birds of Paradise*. London: Collins, 2012.

Audubon, John J. *Birds of America*. New York: Welcome Rain Publishers, 2000 [1827–38].

Barad, Karen. "Posthumanist Performativity: Toward an Understanding of How Matter Comes to Matter." *Journal of Women in Culture and Society* 28 (2003): 801–31. doi: 10.1017/s0140525x98001277.

Bargheer, Stefan. "The Sociology of Morality as Ecology of Mind: Justifications for Conservation and the International Law for the Protection of Birds in Europe." *European Journal of Sociology* 59 (2018): 63–89. https://doi.org/10.1017/S000397561800005X.

Baring-Gould, Sabine, and Alfred Newton. *Iceland: Its Scenes and Sagas*. London: Smith, Elder, and Co., 1863.

Barkham, Patrick. "Half the Trees in Two New English Woodlands Planted by Jays, Study Finds." *Guardian*, 16 June 2021. https://www.theguardian.com/environment/2021/jun/16/half-the-trees-in-two-new-english-woodlands-planted-by-jays-study-finds.

———. "It's Very Easy to Save a Species: How Carl Jones Rescued More Endangered Animals Than Anyone Else." *Guardian*, 26 November 2018. https://www.theguardian.com/environment/2018/nov/26/its-very-easy-to-save-a-species-how-carl-jones-rescued-more-endangered-animals-than-anyone-else.

Barnett, Joshua Trey. *Mourning in the Anthropocene: Ecological Grief and Earthly Coexistence*. East Lansing: Michigan State University Press, 2022.

Barrow, Mark V. *Nature's Ghosts: Confronting Extinction from the Age of Jefferson to the Age of Ecology*. Chicago: University of Chicago Press, 2009.

Barrowclough, George F., Joel Cracraft, John Klicka, and Robert M. Zink. "How Many Kinds of Birds Are There and Why Does It Matter?" *PLOS ONE*, 23 November 2016: 1–15. https://doi.org/10.1371/journal.pone.0166307.

Beer, Gillian. "Darwin and the Uses of Extinction." *Victorian Studies* 51, no. 2 (2009): 321–31. https://www.jstor.org/stable/i20537400.

Bengtson, Sven-Axel. "Breeding Ecology and Extinction of the Great Auk (*Pinguinus impennis*): Anecdotal Evidence and Conjectures." *The Auk, a Quarterly Journal of Ornithology* 101 (1984): 1–11. https://doi.org/10.1093/auk/101.1.1.

Bennett, John. *The Ecological Transition: Cultural Ecology and Human Adaptation*. New York: Pergamon Press, 1976.

Birkhead, Tim R. *Birds and Us: A 12,000-Year History, from Cave Art to Conservation.* New York: Penguin, 2022.

———. "The Chick-Rearing Period of the Great Auk: A Mystery Solved." *British Birds* 114 (2021): 1–62.

———. *Great Auk Islands: A Field Biologist in the Arctic.* London: Poyser Ltd., 1993.

Birkhead, Tim R., Graham Axon, and James R. Middleton. "Restoration of Two Great Auk (*Pinguinus impennis*) Eggs: Bourman Labrey's Eggs and the Scarborough Egg." *Archives of Natural History* 47 (2020): 392–401. https://doi.org/10.3366/anh.2020.0663.

Birkhead, Tim R., Jürgen Fiebig, Robert Montgomerie, and Karl Schulze-Hagen. "The Great Auk (*Pinguinus impennis*) Had Two Brood Patches, Not One; Confirmation and Implications." *Ibis*, 21 September 2021. https://doi.org/10.1111/ibi.13019.

Birkhead, Tim R., and Peter T. Gallivan. "Alfred Newton's Contributions to Ornithology: A Conservative Quest for Facts Rather than Grand Theories." *Ibis*, 18 September 2012. https://doi.org/10.1111/j.1474-919X.2012.01274.x.

Blencowe, Michael. *Gone: A Search for What Remains of the World's Extinct Creatures.* Brighton: Leaping Hare Press, 2021.

Bourne, W.R.P. "The Story of the Great Auk *Pinguinus impennis*." *Archives of Natural History* 20 (1993): 257–78. https://doi.org/10.3366/anh.1993.20.2.257.

Bragadóttir, Kristín. "Fugla- og náttúruvernd Nielsens kaupmanns." In *Bakkadrottningin Eugenía Nielsen*, 328–31. Reykjavík: Ugla, 2022.

Braun, Ingmar M. "Representations of Birds in the Eurasian Upper Palaeolithic Ice Age Art." *Boletim do Centro Português de Geo-História e Pré-História* 1 (2018): 13–21.

Brogaard, Berit. "Species as Individuals." *Biology and Philosophy* 19 (2004): 222–42. https://doi.org/10.1023/B:BIPH.0000024322.46358.61.

Broughton, Richard K., James M. Bullock, Charles George, et al. "Long-Term Woodland Restoration on Lowland Farmland through Passive Rewilding." *PLOS ONE* 16 June 2021. https://doi.org/10.1371/journal.pone.0252466.

Brown, Nancy Marie. *Looking for the Hidden Folk: How Iceland's Elves Can Save the Earth.* New York: Pegasus Books, 2022.

Browne, J. Ross. *The Land of Thor.* New York: Harper, 1867.

Brunner, Bernd. *The Ocean at Home: An Illustrated History of the Aquarium.* London: Reaction Books, 2011.

Buranyi, Stephen. "Do We Need a New Theory of Evolution?" *Guardian*, 28 June 2022, 1–13. https://www.theguardian.com/science/2022/jun/28/do-we-need-a-new-theory-of-evolution.

Butler, Judith. *Precarious Life: The Powers of Mourning and Violence.* London: Verso, 2004.

Canguilhem, Georges. *Knowledge of Life.* New York: Fordham University Press, 2008 [1965].

Chernela, Janet. "A Species Apart: Ideology, Science, and the End of Life." In *The Anthropology of Extinction: Essays on Culture and Species Death*, edited by Genese Marie Sodikoff, 18–38. Bloomington and Indianapolis: Indiana University Press, 2012.

Cherry-Garrard, Apsley. *The Worst Journey in the World*. New York: Skyhorse Publishing, 2016 [1922].

Clottes, Jean. *What Is Paleolithic Art? Cave Paintings and the Dawn of Human Creativity*. Chicago: University of Chicago Press, 2016.

Clottes, Jean, Jean Courtin, and Luc Vanrell. *Cosquer Redécouvert*. Paris: Éditions du Seuil, 2015.

Cole, Edward. "Blown Out: The Science and Enthusiasm of Egg Collecting in the *Oologists' Record*, 1921–1969." *Journal of Historical Geography* 51 (2016): 19–28. https://doi.org/10.1016/j.jhg.2015.10.014.

Collett, Robert. "En Vintergjæst." *Naturen: Et illustreret Manedsskrift for populær Naturvidenskab* 3 (1877): 33–38.

Connolly, William. *Facing the Planetary: Entangled Humanism and the Politics of Swarming*. London: Duke University Press, 2017.

Costa, James T. "Wallace, Darwin, and Natural Selection." In *An Alfred Russel Wallace Companion*, edited by Charles H. Smith, James T. Costa, and David Collard, 97–143. Chicago: University of Chicago Press, 2019.

Cowles, Henry M. "A Victorian Extinction: Alfred Newton and the Evolution of Animal Protection." *British Journal for the History of Science* 46 (2013): 695–714.

Crair, Ben. "The Bizarre Bird That's Breaking the Tree of Life." *New Yorker*, 15 July 2022. https://www.newyorker.com/science/elements/the-bizarre-bird-thats-breaking-the-tree-of-life.

Cunich, Peter, David Hoyle, Eamon Duffy, and Ronald Hyam. *A History of Magdalene College Cambridge, 1428–1988*. Cambridge: Magdalene College Publications, 1994.

Danell, Kjell. "Äggherren och hans runsten i Muoniovaara." *Oknytt Årsbok*, 2022, 97–102.

Darwin, Charles. *The Descent of Man, and Selection in Relation to Sex*. London: John Murray, 1871.

———. *On the Origin of Species by Means of Natural Selection*. Edited with an introduction by J. W. Burrow. London: Penguin, 1968 [1859].

Daston, Lorraine, and Peter Galison. *Objectivity*. Princeton, NJ: Princeton University Press, 2010.

Davies, Tammy E., Ana P. B. Carneiro, Marguerite Tarzia, et al. "Multispecies Tracking Reveals a Major Seabird Hotspot in the North Atlantic." *Conservation Letters*, July 2021, 1–14. doi: 10.1111/conl.12824.

de Buffon, Georges-Louis Leclerc le Comte. *The Epochs of Nature*. Edited by Jan Zalasiewicz, Sophie Milon, and Mateusz Zalasiewicz. Introduction by Jan Zalasiewicz, Sverker Sörlin, Libby Robin, and Jacques Grinevald. Chicago: University of Chicago Press, 2018 [1778].

Delisle, Richard D. *Charles Darwin's Incomplete Revolution: The* Origin of Species *and the Static Worldview*. Lethbridge, Alberta: Springer, 2019.

DeLoughrey, Elizabeth M. *Allegories of the Anthropocene*. Durham, NC: Duke University Press, 2019.

de Lozoya, Arturo Valledor, David Gonzáles García, and Jolyon Parish. "A Great Auk for the Sun King." *Archives of Natural History* 43 (2016): 41–56. https://doi.org/10.3366/anh.2016.0345.

Despret, Vinciane. *The Dance of the Arabian Babbler: Birth of an Ethological Theory*. Transl. Jeffrey Bussolini. Minneapolis: University of Minnesota Press, 2021.

———. "Inhabiting the Phonocene with Birds." In *Critical Zones: The Science and Politics of Landing on Earth*, edited by Bruno Latour and Peter Weibel, 254–59. Cambridge, MA: MIT Press, 2020.

———. *Living as a Bird*. Transl. Helen Morrison. London: Polity Press, 2021.

Dupré, John. *Processes of Life: Essays in the Philosophy of Biology*. Oxford: Oxford University Press, 2012.

———. "(Some) Species Are Processes." In *Species Problems and Beyond: Contemporary Issues in Philosophy and Practice*, edited by John Wilkins, Franz Zachos, and Igor Pavlinov, 278–92. Boca Raton, FL: CRC Press, 2022.

Duyker, Edward. *Nature's Argonaut: Daniel Solander 1733–1782*. Melbourne: Miegunyah Press, 1998.

Eastham, Anne, and Michael Eastham. "Palaeolithic Images and the Great Auk." *Antiquity* 69 (1995): 1023–25. https://doi.org/10.1017/S0003598X00082582.

Ehrlich, Paul, and Ann Ehrlich. *Extinction: The Causes and Consequences of the Disappearance of Species*. New York: Random House, 1981.

Einarsson, Stefán. *Saga Eiríks Magnússonar í Cambridge*. Reykjavík: Ísafoldarprentsmiðja, 1933.

Einarsson, Þorsteinn. "Það var hin mesta hættuför." *Vikan* 25 (1963): 22–23, 51.

Ekman, Mattias. "The Birth of the Museum in the Nordic Countries: *Kunstkammer*, Museology and Museography." *Nordic Museology* 1 (2018): 5–6. https://doi.org/10.5617/nm.6395.

Emling, Shelley. "Mary Anning: Fossil Hunter." *Reports of the National Center for Science Education* 30, no. 3 (2010): 1–10. https://ncse.ngo/mary-anning-fossil-hunter.

Endersby, Jim. "Sympathetic Science: Charles Darwin, Joseph Hooker, and the Passions of Victorian Naturalists." *Victorian Studies* 51 (2009): 299–320. https://www.jstor.org/stable/20537406.

Evans, Kate. "From Healthy to Extinct in 350 Years." *Hakai Magazine*, 6 January 2020. https://www.hakaimagazine.com/news/from-healthy-to-extinct-in-350-years/.

Faber, Frederik. "Beyträge zur arctischen Zoologie." *Isis (von Oken)* 8 (1827): 678–88.

Falk, Oren. *Violence and Risk in Medieval Iceland: This Spattered Isle*. Oxford: Oxford University Press, 2021.

Feeley-Harnik, Gillian. "The Geography of Descent." *Proceedings of the British Academy* 125 (2004): 311–64. doi: 10.5871/bacad/9780197263242.003.0013.

Finlayson, Clive. *The Smart Neanderthal: Bird Catching, Cave Art and the Cognitive Revolution*. Oxford: Oxford University Press, 2019.

Fletcher, Amy Lynn. *De-extinction and the Genomics Revolution*. London: Palgrave Macmillan, 2020.

Frjáls þjóð. "Brúðarrán á Suðurnesjum," 26 January 1957.

Fuentes, Augustine. "'The Descent of Man,' 150 Years On." *Science* 372 (2021): 769. doi: 10.1126/science.abj4606.

Fuller, Errol. *Dodo: From Extinction to Icon*. London: Collins, 2002.

———. *Extinct Birds*. Revised edition. Ithaca, NY: Comstock Publishing Associates, 2001.

———. *The Great Auk*. New York: Harry N. Adams, 1999.

———. *The Great Auk: The Extinction of the Original Penguin*. Chicago: Bunker Hill Publishing, 2003.

———. *The Passenger Pigeon*. Princeton, NJ: Princeton University Press, 2014.

Galton, David. "Did Darwin Read Mendel?" *QJM* 102 (2009): 587–89. https://doi.org/10.1093/qjmed/hcp024.

Gamble, Clive. *Making Deep History: Zeal, Perseverance, and the Time Revolution of 1859*. Oxford: Oxford University Press, 2021.

Garlick, Ben, and Kate Symons. "Geographies of Extinction: Exploring the Spatiotemporal Relations of Species Death." *Environmental Humanities* 12, no. 1 (2020): 296–320.

Gaskell, Jeremy. *Who Killed the Great Auk?* Oxford: Oxford University Press, 2000.

Goldfarb, Ben. *Eager: The Surprising Life of Beavers and Why They Matter*. London: Chelsea Green Publishing, 2018.

Gosse, Philip Henry. *The Aquarium: An Unveiling of the Wonders of the Deep Sea*. London: John Van Voorst, 1854.

Granata, Silvia. *The Victorian Aquarium: Literary Discussions on Nature, Culture, and Science*. Manchester: Manchester University Press, 2021.

Greene, John. *The Death of Adam: Evolution and Its Impact on Western Thought*. Ames: Iowa State University, 1959.

Greengrass, Jessie. *An Account of the Decline of the Great Auk, According to One Who Saw It*. London: John Murray, 2015.

Greenlaw, Lavinia. *Questions on Travel: William Morris in Iceland*. Devon: Notting Hill Editions, 2011.

Grieve, Symington. *The Great Auk, or Garefowl*. Cambridge: Cambridge University Press, 2015 [1885].

Grimbert, Sibylle. *Le Dernier des siens*. Paris: Éditions Anne Carrière, 2022.

Guardian. "Cassowary Attack: Giant Bird Kills Owner in Florida after He Fell," 14 April 2019. https://www.theguardian.com/us-news/2019/apr/14/cassowary -attack-giant-bird-kills-owner-in-florida-after-he-fell.

Guðmundsson, Brandur. "Lýsing á Höfnum." In *Blanda: Fróðleikur gamall og nýr*. Sögurit XVII. Reykjavík: Sögufélag, 1921–23 [1840].

Guðmundsson, Gils. *Íslenzkir athafnamenn I: Geir Zoëga kaupmaður og útgerðar- maður*. Akranes: Akranesútgáfan, 1946.

Hagalín, Guðmundur Gíslason. *Saga Eldeyjar-Hjalta. Skráð eftir frásögn hans sjálfs*. Reykjavík: Almenna bókafélagið, 1974.

Hannah, Michael. *Extinctions: Living and Dying in the Margin of Error*. Cambridge: Cambridge University Press, 2021.

Haraway, Donna. *Staying with the Trouble: Making Kin in the Chthulucene*. Durham, NC: Duke University Press, 2016.

———. "Teddy Bear Patriarchy: Taxidermy in the Garden of Eden, New York City, 1908–1936." In *Primate Visions: Gender, Race, and Nature in the World of Modern Science*. London: Routledge, 1989.

———. *When Species Meet*. Minneapolis: University of Minnesota Press, 2008.

Hastrup, Kirsten. *Nature and Policy in Iceland 1400–1800: An Anthropological Analysis of History and Mentality*. Oxford: Clarendon Press, 1990.

Headley, Maria Dahvana. "Introduction." *Beowulf: A New Translation*. Melbourne: Scribe, 2020.

Heaton, Trevor. "We're Still Fighting Alfred's Battles 150 Years On." *Eastern Daily Press*, 9 October 2020. https://www.edp24.co.uk/lifestyle/conservation-law -inspired-by-norwich-speech-1234138.

Helgason, Jón. "Reykjavíkurkaupmenn á æskuárum mínum." *Frjáls verslun* 2 (1940): 4–6, 31.

Hill, Erica. "Archaeology and Animal Persons: Toward a Prehistory of Human- Animal Relations." *Environment and Society: Advances in Research* 4 (2013): 117– 36. https://doi.org/10.3167/ares.2013.040108.

Hoare, Philip. "Under the Skin of the Ocean, There's a Super-Loud Fishcotheque Going On." *Guardian*, 9 December 2021. https://www.theguardian.com /commentisfree/2021/dec/09/ocean-loud-fish-indonesian-reef-sound.

Hogenboom, Melissa. "How Did the Last Neanderthals Live?" *BBC Future*, 29 Janu- ary 2021. https://www.bbc.com/future/article/20200128-how-did-the-last -neanderthals-live.

Houston, A. I., J. Wood, and M. Wilkinson. "How Did the Great Auk Raise Its Young?" *Evolutionary Biology* 23 (2010): 1899–1906. doi: 10.1111/j.1420–9101.2010 .02047.x.

Hume, J. P., A. S. Cheke, and A. McOran-Campbell. "How Owen 'Stole' the Dodo: Academic Rivalry and Disputed Rights to a Newly-Discovered Subfossil Deposit in Nineteenth Century Mauritius." *Historical Biology* 21 (2009): 33–49. https://doi.org/10.1080/08912960903101868.

Ingold, Tim. *Evolution and Social Life*. New York: Routledge, 1986.

———. "On Extinction." *Clearing*, 29 November 2018. https://www.littletoller.co.uk /the-clearing/on-extinction-by-tim-ingold/.

———. *Imagining for Real: Essays on Creation, Attention and Correspondence*. London: Routledge, 2022.

———. *The Rise and Fall of Generation Now*. Oxford: Polity, 2023.

Ingold, Tim, and Gísli Pálsson (eds.). *Biosocial Becomings: Integrating Social and Biological Anthropology*. Cambridge: Cambridge University Press, 2013.

Ísaksson, Sigurjón Páll. *Úr sögu geirfuglsins á Íslandi*. Reykjavík: Published by the author, 2020.

Jakubas, Darius, Katarzyba Wojczulanis-Jakubas, and Ævar Petersen. "A Quiet Extirpation of the Breeding Little Auk *Alle alle* Population in the Shadow of the Famous Cousin Extermination." *Science of the Total Environment* 808 (2022): 1–12. https://doi.org/10.1016/j.scitotenv.2021.152167.

Johnson, Curtis N. "Charles Darwin, Richard Owen, and Natural Selection: A Question of Priority." *Journal of the History of Biology* 52 (2019): 45–85. doi: 10.1007/s10739 -018-9514-2.

Jónasson, Finnur, Sólveig Ólafsdóttir, and Sigurður Gylfi Magnússon. *Híbýli fátæktar: Húsnæði og veraldleg gæði fátæks fólks á 19. og fram á 20. öld*. Reykjavík: Háskólaútgáfan, 2019.

Jónsson, Finnur. *Þjóðhættir og ævisögur frá 19. öld: Minnisblöð Finns á Kjörseyri*. Akureyri: Bókaútgáfa Pálma H. Jónssonar, 1945.

Jónsson, Jónas. *Dýrafræði*, vol. 2. Reykjavík: Ríkisútgáfa námsbóka, 1945.

Jørgensen, Dolly. "Endling, the Power of the Last in an Extinction-Prone World." *Environmental Philosophy* 14 (2017): 119–38. https://www.jstor.org/stable/2689 4342.

———. *Recovering Lost Species in the Modern Age: Histories of Longing and Belonging*. Cambridge, MA: MIT Press, 2019.

Kalshoven, Petra Tjitske. "Piecing Together the Extinct Great Auk: Techniques and Charms of Contiguity." *Environmental Humanities* 10 (2018): 150–70. https://doi .org/10.1215/22011919-4385507.

Karlsson, Gunnar. *A Brief History of Iceland*. Translation from the Icelandic by Anna Yates. Reykjavík: Mál og menning, 2012.

Ketilsson, Ólafur. "Geirfugl, örn og valur." *Vísir*, 4 October 1929.

———. "Síðustu geirfuglarnir." *Vísir*, 21 July 1929.

Kingsley, Charles. *The Water-Babies*. Nottinghamshire: Award Publications, 2003 [1863].

Kjartansdóttir, Katla. "The Changing Symbolic Meaning of the Extinct Great Auk and Its Afterlife as a Museum Object at the Natural History Museum of Denmark." *Nordisk Museologi* 2 (2019): 41–56. https://doi.org/10.5617/nm.7476.

Koerner, Lisbet. *Linnaeus: Nature and Nation*. Cambridge, MA: Harvard University Press, 1999.

Kolbert, Elizabeth. *The Sixth Extinction: An Unnatural History*. New York: Henry Holt, 2014.

Koons, Ryan A. "Becoming Avian in the Anthropocene: Performing the Feather Dance and the Owl Dance at Pvlvcekolv." *HUMaNIMALIA* 10, no. 2 (2019): 95–127. https://doi.org/10.52537/humanimalia.9503.

Krebber, André, and Mieke Roscher. "Introduction: Biographies, Animals and Individuality." In *Animal Biography: Re-framing Animal Lives*, edited by André Krebber and Mieke Roscher, 1–15. Cham, Switzerland: Palgrave, 2018.

Kristjánsson, Lúðvík. "Geirfugl." In *Íslenzkir sjávarhættir*, vol. 5, 263–65. Reykjavík: Menningarsjóður, 1984.

Lambert, Robert A. "From Exploitation to Extinction, to Environmental Icon." In *Species History in Scotland: Introductions and Extinctions since the Ice Age*, edited by Robert A. Lambert, 20–37. Edinburgh: Scottish Cultural Press, 1998.

Latour, Bruno, and Peter Weibel (eds.). *Critical Zones: The Science and Politics of Landing on Earth*. Cambridge, MA: MIT Press, 2020.

Le Quellec, Jean-Loïc. *La caverne originelle: Art, mythes et premières humanités*. Paris: La Découverte, 2022.

Lestel, Dominique, Jeffrey Bussolini, and Matthew Chrulew. "The Phenomenology of Animal Life." *Environmental Humanities* 5 (2014): 125–48. https://doi.org/10.1215/22011919-3615442.

Lewontin, Richard. "Gene, Organism, and Environment." In *Cycles of Contingency: Developmental Systems and Evolution*, edited by Susan Oyama, Paul E. Griffiths, and Russell D. Gray, 59–66. Cambridge, MA: MIT Press, 2001.

Lien, Marianne Elisabeth, and Gísli Pálsson. "Ethnography beyond the Human: The 'Other-Than-Human' in Ethnographic Work." *Ethnos* 86 (2021): 1–20. https://doi.org/10.1080/00141844.2019.1628796.

Lock, Margaret, and Gísli Pálsson. *Can Science Resolve the Nature/Nurture Debate?* Oxford: Polity, 2016.

Lomolina, Mark V. "Wallace at the Foundations of Biogeography and the Frontiers of Conservation Biology." In *An Alfred Russel Wallace Companion*, edited by Charles H. Smith, James T. Costa, and David Collard, 341–54. Chicago: University of Chicago Press, 2019.

Lorimer, Jamie. "On Auks and Awkwardness." *Environmental Humanities* 4 (2014): 195–205. https://doi.org/10.1215/22011919-3614989.

Love, John A. "The Last Great Auk." *Scottish Birds* 31 (2011): 341–47.

Lund, Katrín Anna, Katla Kjartansdóttir, and Kristín Loftsdóttir "'Puffin Love': Performing and Creating Arctic Landscapes in Iceland through Souvenirs." *Tourist Studies* 18 (2018): 142–58. https://doi.org/10.1177/1468797617722353.

Magnússon, Árni, and Páll Vídalín. *Jarðabók*, vol. 3. Copenhagen: S. L. Möller, 1923–24.

Marguilies, Jared. "A Political Ecology of Desire: Between Extinction, Anxiety, and Flourishing." *Environmental Humanities* 14, no. 2 (2022): 241–64. https://doi.org/10.1215/22011919-9712357.

Marsh, George Perkins. *Man and Nature: On Physical Geography as Modified by Human Action.* New York: Charles Scribner & Co., 1867.

Martin, Martin. *A Late Voyage to St. Kilda, the Remotest of the Hebrides, or Western Isles of Scotland.* London: Gent, 1698.

Maurer, Konrad. *Íslandsferð 1858.* Trans. by Baldur Hafstað. Reykjavík: Ferðafélag Íslands, 1997.

Meldgaard, Morten. "The Great Auk, *Pinguinus impennis* (L.) in Greenland." *Historical Biology* 1 (1988): 145–78. https://doi.org/10.1080/08912968809386472.

Meloni, Maurizio, Rachael Wakefield-Rann, and Becky Mansfield. "Bodies of the Anthropocene." *Anthropocene Review* (2021): 1–21. https://doi.org/10.1177/2053019621100517.

Mishler, Brent D. "Ecology, Evolution, and Systematics in a Post-Species World." In *Species Problems and Beyond: Contemporary Issues in Philosophy and Practice*, edited by John Wilkins, Franz Zachos, and Igor Pavlinov, 177–90. Boca Raton, FL: CRC Press, 2022.

Monbiot, George. *Feral: Rewilding the Land, Sea and Human Life.* London: Penguin, 2013.

Montgomery, Sy. *The Hummingbirds' Gift: Wonder, Beauty, and Renewal on Wings.* New York: Astria Books.

Morgunblaðið. "Gríðarlega mikil plastmengun í Eldey," 8 December 2022, 2.

Morris, John G. L., and Andrew Duggins. "The Avian Blink: The Phenomenon of Blinking on Head Turns." *Winks and Blinks: A Knowledge Platform*, 2019. https://www.winks-and-blinks.com/blinks-on-head-turns.

Morris, William. *Dagbækur úr Íslandsferðum 1871–1873.* Reykjavík: Mál og menning, 1975 [1910–15].

Moynihan, Thomas. "The End of Us." *Aeon*, 7 August 2019. https://aeon.co/essays/to-imagine-our-own-extinction-is-to-be-able-to-answer-for-it.

Müller-Wille, Staffan, and Karen Reeds. "A Translation of Carl Linnaeus's Introduction to *Genera plantarum* (1737)." *Studies in History and Philosophy of Biological and Biomedical Science* 38 (2007): 563–72.

Müller-Wille, Staffan, and Hans-Jörg Rheinberger (eds.). *Heredity Produced: At the Crossroads of Biology, Politics, and Culture, 1500–1870.* Cambridge, MA: MIT Press, 2007.

Mynott, Jeremy. *Birds in the Ancient World: Winged Words.* Oxford: Oxford University Press, 2018.

Nathorst, A. G. "Carl von Linné as a Geologist." *Annual Report of the Smithsonian Institution*, 1908.

Naturhistorisk tidskrift. "The Great Auk in Iceland?" (1838–39), 207.

Newton, Alfred. "Abstract of Mr. J. Wolley's Researches in Iceland Respecting the Gare-Fowl or Great Auk (*Alca impennis*, Linn)." *Ibis* (1861): 374–99.

———. "Address to the Department of Botany and Zoology." In *Report of the Forty-Fifth Annual Meeting of the British Association for the Advancement of Science,* 119–25. London: John Murray, 1876.

———. "Appendix A: Notes on the Ornithology of Iceland." In Sabine Baring-Gould and Alfred Newton, *Iceland: Its Scenes and Sagas,* 399–421. London: Smith, Elder, and Co., 1863.

———. "Cuckoo's Eggs." *Nature* 1 (1869): 74–76.

———. *A Dictionary of Birds.* Assisted by Hans Gadow. London: Adam and Charles Black, 1896.

———. "On Existing Remains of the Gare-fowl (Alca impennis)." *Ibis* (1870): 257–61. https://doi.org/10.1111/j.1474-919X.1870.tb05797.x.

———. "The Gare-Fowl and Its Historians." *Natural History Review* 5 (1865): 467–88.

———. "The Great Auk, or Garefowl (*Alca impennis*, Linn), Its History, Archaeology, and Remains." *Nature* 32 (1885): 545–46.

———. "Memoir." In John Wolley and Alfred Newton, 1864–1907, *Ootheca Wolleyana: An Illustrated Catalogue of the Collection of Birds' Eggs Begun by the Late John Wolley,* edited by Alfred Newton, ix–xxxix. London: R. H. Porter, 1864.

———. "Memoir of the Late John Wolley." *Ibis* (1860): 175–85.

———. "On the Possibility of Taking an Ornithological Census." *Ibis* (1861): 190–96.

———. "On the Zoological Aspects of Game Laws." In *Report of the Thirty-Eighth Meeting of the British Association for the Advancement of Science; Held at Norwich in August 1868,* 107–8. Norwich: British Association for the Advancement of Science, 1869.

Newton, Alfred, and Edward Newton. "Observations on the Birds of St. Croix, West Indies." *Ibis* (1859): 138–50.

Nice, Margaret Morse. "The Role of Territory in Bird Life." *American Midland Naturalist* 26, no. 3 (1941): 441–47. https://doi.org/10.2307/2420732.

Nicolson, Adam. *The Seabird's Cry: The Lives and Loves of the Planet's Great Ocean Voyagers.* New York: Henry Holt, 2018.

Nielsen, Peter. "Geirfugl, sæörn og fálki: Þrír alíslenskir fuglar." *Lesbók Morgunblaðsins* 19, 12 May 1929.

———. "Síðustu geirfuglarnir." *Vísir*, 12 September 1929.

Nijhuis, Michelle. *Beloved Beasts: Fighting for Life in the Age of Extinction*. New York: W. W. Norton, 2022.

Nordén, Åsa. *Sällsamheter i Tornedalen*. Stockholm: Rabén & Sjögren, 1983.

Ólafsson, Eggert, and Bjarni Pálsson. *Ferðabók Eggerts Ólafssonar og Bjarna Pálssonar um ferðir þeirra á Íslandi árin 1752–1757*. 2 vols. Reykjavík: Örn og Örlygur, 1974 [1942].

———. *Rejse igiennem Island*. Söröe: Det Kongelige danske Videnskabernes Selskab, 1772.

Ólason, Páll Eggert. *Íslenskar æviskrár frá landnámstímum til ársloka 1940*. Vol. 6. Reykjavík: Hið íslenzka bókmenntafélag, 1948–76.

Olson, Storrs L., and Judith N. Lund. "Whalers and Woggins: A New Vocabulary for Interpreting Some Early Accounts of the Great Auk and Penguins." *Archives of Natural History* 34 (2007): 69–77. https://doi.org/10.3366/anh.2007.34.1.69.

Óskarsson, Þorleifur. *Saga Reykjavíkur—í þúsund ár: 870–1870*. Reykjavík: Iðunn, 2002.

Owen, Richard. "On the Extinction of Species." In *On the Classification and Geographical Distribution of the Mammalia*. London: J. W. Parker & Son, 1959, 56, Appendix A.

———. *Palæontology or a Systematic Summary of Extinct Animals and Their Geological Relations*. Edinburgh: Adam and Charles Black, 1860.

Palmbo, Åsa. "Äggherren och runristningen i Muoniovaara." *Kulturmiljö vid Norrbottens Museum*, 13 July 2018, 1–12.

Pálsson, Gísli. "Biosocial Relations of Production." *Comparative Studies in Society and History* 51 (2009): 288–313. https://www.jstor.org/stable/40270328.

———. "The Birth of the Aquarium: The Political Ecology of Icelandic Fishing." In *The Politics of Fishing*, edited by T. Gray, 209–27. London: Macmillan, 1998.

———. *Coastal Economies, Cultural Accounts: Human Ecology and Icelandic Discourse*. Manchester: Manchester University Press, 1991.

———. "Ensembles of Biosocial Relations." In *Biosocial Becomings: Integrating Social and Biological Anthropology*, edited by Tim Ingold and Gísli Pálsson, 22–41. Cambridge: Cambridge University Press, 2013.

———. "Enskilment at Sea." *Man*, New Series 29 (1994): 901–27. https://doi.org/10.2307/3033974.

———. *The Human Age: How We Created the Anthropocene Epoch and Caused the Climate Crisis*. London: Welbeck, 2020.

———. *The Man Who Stole Himself: The Slave Odyssey of Hans Jonathan*. Chicago: University of Chicago Press, 2016.

———. "Vegabréf fuglanna: Fuglamerkingar Óskars J. Sigurðssonar." *Náttúrufræðingurinn*, December 2022.

Pálsson, Gísli, and E. Paul Durrenberger. "Icelandic Foremen and Skippers: The Structure and Evolution of a Folk Model." *American Ethnologist* 10 (1983): 511–28. https://www.jstor.org/stable/644267.

Papagianni, Dimitra, and Michael A. Morse. *The Neanderthals Rediscovered: How Modern Science Is Rewriting Their Story.* London: Thames & Hudson, 2022.

Parkin, Simon. "Darwin's Lost Treasure, Found." *New Yorker*, 16 August 2022.

Pietsch, Theodore W. *Trees of Life: A Visual History of Evolution.* Baltimore, MD: Johns Hopkins University Press, 2012.

Pope, Peter E. "Early Migratory Fishermen and Newfoundland's Seabird Colonies." *Journal of the North Atlantic* 1 (2009): 57–74. https://www.jstor.org/stable/2668 6926.

Porr, Martin. "Rock Art as Art." *Time and Mind* 12, no. 2 (2019): 153–64.

Posnett, Edward. "The Weird Magic of Eiderdown." *Guardian*, 19 July 2019. https://www .theguardian.com/world/2019/jul/19/eiderdown-harvesting-iceland-eider-duck.

Purchell, Rosamond Wolff, and Stephen Jay Gould. *Finders, Keepers: Eight Collectors.* London: Hutchinson, 1993.

Pyenson, Nick. *Spying on Whales.* London: William Collins, 2019.

Pyne, Lydia. *Endling: Fables for the Anthropocene.* Minneapolis: University of Minnesota Press, 2022.

———. "What a New Translation of *Beowulf* Says about Extinction." *Literary Hub*, 18 August 2022. https://lithub.com/what-a-new-translation-of-beowulf-says-about -extinction/.

Quammen, David. *The Song of the Dodo: Island Biogeography in the Age of Extinctions.* New York: Scribner, 2011.

Raffles, Hugh. *The Book of Unconformities: Speculations on Lost Time.* New York: Pantheon Books, 2020.

Reinert, Hugo. "The Haunting Cliffs: Some Notes on Silence." *Field Philosophy and Other Experiments* 24 (2018): 501–12. https://doi.org/10.1080/13534645.2018 .1546726.

Reisz, Matthew. "Great Auks and Seal-Headed Men: A Window into Ice Age Provence." *Observer*, 2 July 2022. https://www.theguardian.com/science/2022/jul /02/ice-age-provence-cosquer-cave-archaeology.

Rheinberger, Hans-Jörg. "Gene Concepts." In *The Concept of the Gene in Development and Evolution: Historical and Epistemological Perspectives*, edited by P. J. Beurton, R. Falk, and H. J. Rheinberger, 219–39. Cambridge: Cambridge University Press, 2000.

Rhodes, Richard. *John James Audubon: The Making of an American.* New York: Alfred A. Knopf, 2005.

Richardson, Susan. *Where the Seals Sing: Exploring the Hidden Lives of Britain's Grey Seals.* London: William Collins, 2022.

Richmond, Douglas J., Mikkel-Holger S. Sinding, and M. Thomas P. Gilbert. "The Potential and Pitfalls of De-extinction." *Zoologica Scripta* 45 (2016), S1: 22–36. https://doi.org/10.1111/zsc.12212.

Rieppel, Olivier. "Species as a Process." *Acta Biotheoretica* 57 (2009): 33–49. https://doi.org/10.1007/s10441-008-9057-6.

Rose, Deborah Bird. "Double Death." *The Multispecies Saloon* (n.d.). https://www.multispecies-salon.org/double-death/.

———. *Shimmer: Flying Fox Exuberance in Worlds of Peril.* Edinburgh: Edinburgh University Press, 2022.

———. "Slowly—Writing into the Anthropocene." In *Writing Creates Ecology and Ecology Creates Writing,* edited by Martin Harrison, Deborah Bird Rose, Lorraine Shannon, and Kim Satchell. Special issue of *TEXT* 20 (2013): 3–4. https://doi.org/10.52086/001c.28826.

Rose, Deborah Bird, Thom van Dooren, Matthew Chrulew, and Cary Wolfe (eds.). *Extinction Studies: Stories of Time, Death, and Generations.* New York: Columbia University Press, 2017.

Rudwick, Martin J. S. *Earth's Deep History: How It Was Discovered and Why It Matters.* Chicago: University of Chicago Press, 2014.

Rutz, Christian, Matthias-Claudio Loretto, Amanda E. Bates, et al. "Covid-19 Lockdown Allows Researchers to Quantify the Effects of Human Activity on Wildlife." *Nature Ecology and Evolution* 4 (2020): 1156–59. https://doi.org/10.1038/s41559-020-1237-z.

Sæmundsson, Bjarni. "Fiskirannsóknir 1911–1912." *Andvari* 38 (1913): 1–43.

Scales, Helen. "Is It Time to Begin Rewilding the Oceans?" *Guardian,* 4 July 2021. https://www.theguardian.com/environment/2021/jul/04/rewilding-the-seas-overfishing-oceans.

Scheffler, Samuel, and Niko Kolodny. *Death and the Afterlife.* Oxford: Oxford University Press, 2016.

Schepelern, Henrik Ditlev. *Museum Wormianum: Dets Forutsætninger og Tilblivelse.* Copenhagen: Wormianum, 1971.

Schroer, Sara Asu. "Jakob von Uexküll: The Concept of *Umwelt* and Its Potentials for Anthropology beyond the Human." *Ethnos* 86 (2021): 132–52. https://doi.org/10.1080/00141844.2019.1606841.

Sepkoski, David. *Catastrophic Thinking: Extinction and the Value of Diversity from Darwin to the Anthropocene.* Chicago: University of Chicago Press, 2020.

Serjeantson, Dale. "The Great Auk and the Gannet: A Prehistoric Perspective on the Extinction of the Great Auk." *International Journal of Osteoarchaeology,* April 2001. https://doi.org/10.1002/oa.545.

Shubin, Neil. "Extinction in Deep Time: Lessons from the Past." In *Biological Extinction: New Perspectives*, edited by Partha Dasgupta, Peter H. Raven, and Anna L. McIvor, 22–33. Cambridge: Cambridge University Press, 2019.

Sigari, Dario, Ilaria Mazzini, Jacopo Conti, Luca Forti, Giuseppe Lembo, Beniamino Mecozzi, Brunella Muttillo, and Raffaele Sardella. "Birds and Bovids: New Parietal Engravings at the Romanelli Cave, Apulia." *Antiquity* 95, no. 484 (2021): 1387–1404. https://doi.org/10.15184/aqy.2021.128.

Sklair, Leslie (ed.). *The Anthropocene in Global Media: Naturalizing the Risk*. New York: Routledge, 2021.

Smith, Charles H. *Alfred Russel Wallace: An Anthology of His Shorter Writings*. Oxford: Oxford University Press, 1991.

Smocovitis, Vassiliki Betty. *Unifying Biology: The Evolutionary Synthesis and Evolutionary Biology*. Princeton, NJ: Princeton University Press, 2020.

Sodikoff, Genese Marie (ed.). *The Anthropology of Extinction: Essays of Culture and Species Death*. Bloomington: Indiana University Press, 2012.

Speck, Frank G. *Bird Lore of the Northern Indians*. Reprint from Volume VII, *Public Lectures by University of Pennsylvania Faculty, 1919–1920*. Philadelphia: University of Pennsylvania, 1921.

Spray, Thomas. "The Terrible *Njorl's Saga*: Comedic Reimaginings of the Íslendingasögur from the Victorians to the Present Day." In *The Vikings Reimagined: Reception, Recovery, Engagement*, edited by Tom Birkett and Roderick Dale, 107–28. Boston: Walter de Gruyter, 2020.

Stankowski, Sean, and Mark Ravinet. "Defining the Speciation Continuum." *Evolution* 75–76 (2021): 1256–73.

Steenstrup, Johannes Japetus Smith. "Et Bidrag til Geirfuglens Naturhistorie og særligt til Kundskaben om dens tidligere Udbredningskreds." *Foren. Videnskabelige Meddelelser* 3–7 (1857): 1–116.

Steenstrup, Johannes. *Iapetus Steenstrup i ungdomsaarene 1813–1845, en skildring*. Copenhagen: Bianco Lunos Bogtrykker, 1913.

Steffen, Will, Wendy Broadgate, Lisa Deutsch, Owen Ludwig, and Cornelia Ludwig. "The Trajectory of the Anthropocene: The Great Acceleration." *Anthropocene Review* 2, no. 1 (2015): 81–98. doi: 10.1177/2053019614564785.

Thisted, Kirsten. *Således skriver jeg, Aron: Samlede fortællinger af Aron fra Kangeq (1822–1869)*. Nuuk: Atuakkiorfik, 1999.

Thomas, Jessica E. "Evolution and Extinction of the Great Auk: A Palaeogenomic Approach." Doctoral Dissertation. Bangor University and the University of Copenhagen, 2018.

Thomas, Jessica E., Gary R. Carvalho, James Haile, et al. "An 'Aukward' Tale: A Genetic Approach to Discover the Whereabouts of the Last Great Auks." *Genes* 8 (2017): 1–12. doi: 10.3390/genes8060164.

Thomas, Jessica E., Gary R. Carvalho, James Haile, et al. "Demographic Reconstruction from Ancient DNA Supports Rapid Extinction of the Great Auk." *eLife*, 26 November 2019. doi: 10.7554/eLife.47509.

Thomas, Sarah. "Allegorizing Extinction: Humboldt, Darwin and the Valedictory Image." In *Picturing Evolution and Extinction*, edited by Fae Brauer, 1–18. Newcastle: Cambridge Scholars Publishing, 2015.

Thorarensen, Jón. "Suðurnesjaannáll 1844." *Rauðskinna: Sögur og sagnir*. Reykjavík: Ísafoldarprentsmiðja, 1929–61.

———. "Suðurnesjaannáll 1858." *Rauðskinna: Sögur og sagnir*. Reykjavík: Ísafoldarprentsmiðja, 1929–61.

———. "Um Hafnir í gamla daga." In *Frá Suðurnesjum: Frásagnir frá liðinni tíð*. Reykjavík: Félag Suðurnesjamanna í Reykjavík, 1960.

Thorsen, Liv Emma. *Dyrenes by: Hover, klover og klør i Kristiania 1859–1925*. Oslo: Press, 2020.

Torrens, Hugh. "Mary Anning (1799–1847) of Lyme: The Greatest Fossilist the World Ever Knew." *British Journal for the History of Science* 25 (1995): 257–84. https://doi.org /10.1017/S0007087400033161.

Tsing, Anna Lowenhaupt. "The Sociality of Birds: Reflections on Ontological Edge Effects." In *Kin: Thinking with Deborah Bird Rose*, edited by Thom van Dooren and Matthew Chrulew, 15–32. Durham, NC: Duke University Press, 2022.

Tudge, Colin. *In Mendel's Footnotes: An Introduction to the Science and Technologies of Genes and Genetics from the Nineteenth Century to the Twenty-Second*. London: Vintage, 2002.

Unsöld, Markus. "Two and a Half Auks—the History of the Great Auks, *Pinguinus impennis*, at the ZSM (Charadriiformes, Alcidae)." *Spixiana* 37 (2014): 153–59.

Útfararminning dannebrogsmanns Vilhjálms Kristins Hákonarsonar. Reykjavík: Einar Þórðarson, 1872.

van Dooren, Thom. *Flight Ways: Life and Loss at the Edge of Extinction*. New York: Columbia University Press, 2014.

———. *A World in a Shell: Snail Stories for a Time of Extinctions*. Cambridge, MA: MIT Press, 2022.

von Uexküll, Jakob. *A Foray into the Worlds of Animals and Humans with a Theory of Meaning*. Minneapolis: University of Minnesota Press, 2010 [1934].

Walls, Laura Dassow. *Henry David Thoreau: A Life*. Chicago: University of Chicago Press, 2017.

Walters, Michael. "Uses of Egg Collections: Display, Research, Identification, the Historical Aspect." *Journal of Biological Curation* 1 (1994): 29–35. http://www .natsca.org/article/1057.

Watts, Jonathan. "Biodiversity Crisis Is About to Put Humans at Risk, UN Scientists to Warn." *Guardian*, 3 May 2019. https://www.theguardian.com/environment /2019/may/03/climate-crisis-is-about-to-put-humanity-at-risk-un-scientists -warn.

Wawn, Andrew. "'Fast er drukkið og fátt lært': Eiríkur Magnússon, Old Northern Philology, and Victorian Cambridge." H. M. Chadwick Memorial Lectures. Cambridge: Department of Anglo-Saxon, Norse, and Celtic, Cambridge University, 2000.

Weidensaul, Scott. *A World on the Wing: The Global Odyssey of Migratory Birds*. New York: W. W. Norton, 2021.

Weston, Phoebe. "Half of World's Bird Species in Decline as Destruction of Avian Life Intensifies." *Guardian*, 28 September 2022. https://www.theguardian.com /environment/2022/sep/28/nearly-half-worlds-bird-species-in-decline-as -destruction-of-avian-life-intensifies-aoe.

Whitehouse, Andrew. "Listening to Birds in the Anthropocene: The Anxious Semiotics of Sound in a Human-Dominated World." *Environmental Humanities* 6 (2015): 53–71. doi: 10.1215/22011919-3615898.

Wilkins, John S., and Frank E. Zachos (eds.). *Species Problems and Beyond: Contemporary Issues in Philosophy and Practice*. Boca Raton, FL: CRC Press, 2022.

Willson, Margaret. *Seawomen of Iceland: Survival on the Edge*. Seattle: University of Washington Press, 2016.

Wilson, Robert A. "Continuing after Species." In *Species Problems and Beyond: Contemporary Issues in Philosophy and Practice*, edited by John Wilkins, Franz Zachos, and Igor Pavlinov, 343–53. Boca Raton, FL: CRC Press, 2022.

Wollaston, A.F.R. *Life of Alfred Newton*. London: John Murray, 1921.

Wolley, John, and Alfred Newton. *Ootheca Wolleyana: An Illustrated Catalogue of the Collection of Birds' Eggs Begun by John Wolley*. London: R. H. Porter, 1864–1907.

Wood, Harold B. "The History of Bird Banding." *Auk* 02 (1945): 256–65.

Worster, Donald (ed.). *The Ends of the Earth: Perspectives on Environmental History*. Cambridge: Cambridge University Press, 1988.

Wulf, Andrea. *The Invention of Nature: The Adventures of Alexander von Humboldt*. London: John Murray, 2015.

Zimmer, Carl. "Birds Are Vanishing from North America." *New York Times*, 22 September 2019. https://www.nytimes.com/2019/09/19/science/bird-populations -america-canada.html.

Þjóðólfur. "Brezkir ferðamenn á Íslandi 1858." 32 (1858): 130.

Þór, Jón Þ. *Hafnir á Reykjanesi: Saga byggðar og mannlífs í ellefu hundruð ár*. Reykjanesbær: Reykjanesbær, 2003.

Þórarinsson, Magnús. "Eldeyjarfarir fyrir 60 árum." In *Frá Suðurnesjum: Frásagnir frá liðinni tíð*, 173–88. Reykjavík: Félag Suðurnesjamanna í Reykjavík, 1960.

Þórarinsson, Sigurður. "Neðansjávargos við Ísland." *Náttúrufræðingurinn* 2 (1965): 49–96.

Unpublished Sources

Alfred Newton's Papers. Cambridge University Library, MS Add. 9839.

Guðbjörnsson, Sigurður Örn. "Sundreglur prófessors Nachtegalls." Unpublished manuscript.

Interview with Michael Brooke, Cambridge, 14 May 2019.

Interview with Errol Fuller, Tunbridge Wells, 2 December 2018.

Interview with Thomas Gilbert, Copenhagen, 22 March 2019.

Interview with Vífill Oddsson, Reykjavík, 1 April 2019.

Letter from N. Dahlerup, Thorshavn, to John Wolley, December 1858. John Wolley, *The Gare-Fowl Books*, 1858. Cambridge University Library. MS Add. 9839/2/ Book 4. Pp. 26–33.

Letter from Charles Darwin to Alfred Newton, 8 October 1862. A.F.R. Wollaston, *Life of Alfred Newton*, 124–25. London: John Murray, 1921.

Letter from Charles Darwin to Alfred Newton, 29 October 1865. Cambridge University Library. MS Add. 9839/ID/56.

Letters from Philip Henry Gosse to Alfred Newton, 5 and 9 January 1857. Cambridge University Library. MS Add. 9839/IG/383–386.

Letter from Jón Hjaltalín to John Wolley, 4 December 1858. Cambridge University Library. MS Add. 9839/25/142.

Letter from Charles Kingsley to Alfred Newton, 10 March 1870. See A.F.R. Wollaston, *Life of Alfred Newton*. London: John Murray, 1921.

Letter from Eiríkur Magnússon to Alfred Newton, 8 February 1877. Cambridge University Library. MS Add. 9839/IM/303.

Letter from Sheriff Hans Christopher Müller to John Wolley, 28 May 1858. Cambridge University Library. MS Add. 9839/25/82.

Letter from Alfred Newton to Charles Darwin, 10 March 1874. The Darwin Correspondence Project, Cambridge.

Letter from Alfred Newton to Charles Darwin, 21 January 1867. The Darwin Correspondence Project, Cambridge.

Letter from Augusta Plesner to G. D. Rowley, 19 January 1873. John Wolley, *Gare-Fowl Books*. Cambridge University Library. MS Add. 9839/2/3. Pp. 83–86.

Letter from Alfred Russel Wallace to Alfred Newton, 19 February 1865. Natural History Museum, London. Wallace's Letters. WCP4006.

Letter from John Wolley to Alfred Newton, 5 April 1852. Cambridge University Library. MS Add. 9839/W/597.

Letter from John Wolley to Alfred Newton, 24 March 1858. Cambridge University Library. MS Add. 9839/IW/593.

Letter from John Wolley to Alfred Newton, 1 April 1858. Cambridge University Library. MS Add. 9839/IW/595.

Letter from John Wolley to Alfred Newton, 3 September 1858. Cambridge University Library. MS Add. 9839/IW/607.

Letter from John Wolley to Alfred Newton, 8 October 1858. Cambridge University Library. MS Add. 9839/IW/613.

Letter from John Wolley to Edward Newton, 27 July 1859. Cambridge University Library. MS Add. 9839/IW/669.

Letter from Geir Zoëga to John Wolley, 2 June 1859. Cambridge University Library. MS Add. 9839/IW/666.

Manuscript of Jón Bjarnason (about 1791–1861), "Náttúrufræðirit." National and University Library of Iceland. ÍBR 69 4to.

Newton, Alfred. "Testimony before the Select Committee on Wild Birds Protection," 19 June 1873. Cambridge University Library. MS Add. 9839/5/1.

Wolley, John. *Gare-Fowl Books*, 1858. Begun by Wolley; continued after his death by Alfred Newton, with extensive annotations by the latter.

Websites

Ancestry.ca: https://www.ancestry.ca/genealogy/records/richard-bridgman-barrow-24-10g2jt.

Ashworth, James: "HMS Challenger: How a 150-Year-Old Expedition Still Influences Scientific Discoveries Today." Natural History Museum, 6 September 2022. https://www.nhm.ac.uk/discover/news/2022/september/hms-challenger-how-150-year-old-expedition-still-influences-scientific-discoveries-today.html.

Audubon: https://www.audubon.org/magazine/spring-2021/what-do-we-do-about-john-james-audubon.

Bird Life International: http://datazone.birdlife.org/species/factsheet/great-auk-pinguinus-impennis/text.

The Book of Icelanders (Íslendingabók): https://www.islendingabok.is.

Brand, Steward. "Resurrecting the Great Auk." https://blog.longnow.org/02016/02/04/is-the-great-auk-a-candidate-for-de-extinction/.

Darwin Correspondence Project, 27 November 1886: https://www.darwinproject.ac.uk.

"The Extinction of the Great Auk." Audubon: https://johnjames.audubon.org/extinction-great-auk.

The Glyptodon: https://theglyptodon.wordpress.com/2011/11/05/the-great-auks
-last-days/.

Historicracing.com: https://www.historicracing.com/driverDetail.cfm?driverID
=7868.

IPCC/United Nations Report: https://www.ipcc.ch/report/ar6/wg1/.

Legacies of British Slave-Ownership: https://www.ucl.ac.uk/lbs/person/view
/2146644193.

Library of Congress: https://www.loc.gov/item/2004662124/.

Málfarsbankinn: https://malfar.arnastofnun.is/grein/70211.

Matza, Max. "Why US Bird Attacks on Humans Are on Rise." BBC News, 17 July 2019:
https://www.bbc.com/news/world-us-canada-48993220.

Müller, Wolfgang. Séance Vocibus Avium, 2008: https://www.br.de/radio/bayern2
/sendungen/hoerspiel-und-medienkunst/hoerspiel-seance-vogelart-gesang
-rekonstruktion100.html. https://www.discogs.com/de/Wolfgang-M%C3%BCller
-S%C3%A9ance-Vocibus-Avium/release/1573289.

Oxford English Dictionary: https://public.oed.com/how-words-enter-the-OED
/graphic/.

Roberts, Stuart. "Missing Darwin's Notebooks Returned to Cambridge University
Library." Cambridge University Library: https://www.cam.ac.uk/stories
/TreeOfLife.

Sea Birds Preservation Act, 1869: https://vlex.co.uk/vid/birds-preservation-act-1869
-861253239.

Wilkins, Jonathan. "Ten Extinct Animals Have Been Rediscovered." Gizmodo, 2011:
https://io9.gizmodo.com/ten-extinct-animals-that-have-been-rediscovered
-5822783?fbclid=IwAR2mcuCFoqC6Ciov2GVUUwzpwwA8_tzoYDgecoa6y
_Ya6FuZ2xg9dlPq-a8.

INDEX

Page numbers in *italics* indicate figures and tables.